Reviews of Environmental Contamination and Toxicology

VOLUME 109

Reviews of Environmental Contamination and Toxicology

Continuation of Residue Reviews

Editor
George W. Ware

Founding Editor
Francis A. Gunther

VOLUME 109

Springer-Verlag
New York Berlin Heidelberg
London Paris Tokyo

Coordinating Board of Editors

New York: 175 Fifth Avenue, New York, N.Y. 10010, USA
Heidelberg: 6900 Heidelberg 1, Postfach 105 280, West Germany

Library of Congress Catalog Card Number 62-18595.
Printed in the United States of America.

ISSN 0179-5953

ISBN 0-387-96952-7 Springer-Verlag New York Berlin Heidelberg
ISBN 3-540-96952-7 Springer-Verlag Berlin Heidelberg New York

Foreword

Global attention in scientific, industrial, and governmental communities to traces of toxic chemicals in foodstuffs and in both abiotic and biotic environments has justified the present triumvirate of specialized publications in this field: comprehensive reviews, rapidly published progress reports, and archival documentations. These three publications are integrated and scheduled to provide in international communication the coherency essential for nonduplicative and current progress in a field as dynamic and complex as environmental contamination and toxicology. Until now there has been no journal or other publication series reserved exclusively for the diversified literature on "toxic" chemicals in our foods, our feeds, our geographical surroundings, our domestic animals, our wildlife, and ourselves. Around the world immense efforts and many talents have been mobilized to technical and other evaluations of natures, locales, magnitudes, fates, and toxicology of the persisting residues of these chemicals loosed upon the world. Among the sequelae of this broad new emphasis has been an inescapable need for an articulated set of authoritative publications where one could expect to find the latest important world literature produced by this emerging area of science together with documentation of pertinent ancillary legislation.

The research director and the legislative or administrative adviser do not have the time even to scan the large number of technical publications that might contain articles important to current responsibility; these individuals need the background provided by detailed reviews plus an assured awareness of newly developing information, all with minimum time for literature searching. Similarly, the scientist assigned or attracted to a new problem has the requirements of gleaning all literature pertinent to his task, publishing quickly new developments or important new experimental details to inform others of findings that might alter their own efforts, and eventually publishing all his supporting data and conclusions for archival purposes.

The end result of this concern over these chores and responsibilities and with uniform, encompassing, and timely publication outlets in the field of environmental contamination and toxicology is the Springer-Verlag (Heidelberg and New York) triumvirate:

Reviews of Environmental Contamination and Toxicology (Vol. 1 in 1962 as *Residue Reviews* through Vol. 97 in 1986) for basically detailed review articles concerned with any aspects of chemical contaminants, including pesticides, in the total environment with their toxicological considerations and consequences.

Bulletin of Environmental Contamination and Toxicology (Vol. 1 in 1966) for rapid publication of short reports of significant advances and discoveries in the fields of air, soil, water, and food contamination and pollution as well as methodology and other disciplines concerned with the introduction, presence, and effects of toxicants in the total environment.

Archives of Environmental Contamination and Toxicology (Vol. 1 in 1973) for important complete articles emphasizing and describing original experimental or theoretical research work pertaining to the scientific aspects of chemical contaminants in the environment.

Manuscripts for *Reviews* and the *Archives* are in identical formats and are subject to review, by workers in the field, for adequacy and value; manuscripts for the *Bulletin* are also reviewed but are published by photo-offset to provide the latest results without delay. The individual editors of these three publications comprise the joint Coordinating Board of Editors with referral within the Board of manuscripts submitted to one publication but deemed by major emphasis or length more suitable for one of the others.

Coordinating Board of Editors

Preface

Despite attempts by the media to convince us our surroundings are under continual chemical assault and not faring well, there is abundant evidence that most chemicals are degraded or dissipated in our not-so-fragile environment. Yet, we must contend with leaking underground fuel tanks, movement of nitrates and nitrites into our groundwater reservoirs, increasing air pollution in our large cities, and seemingly frequent contamination of our food and animal feeds with pesticides, industrial chemicals, and bacterial toxins.

Without continuing surveillance and intelligent controls, some of these chemicals could at times conceivably endanger the environment, wildlife, and the public health. Ensuring safety-in-use of the many chemicals involved in our highly industrialized culture is a dynamic challenge, for the old established materials are continually being displaced by newly developed molecules more acceptable to environmentalists, toxicologists, and federal and state regulatory agencies.

These matters are of genuine concern to increasing numbers of governmental agencies and legislative bodies around the world, for some of these chemicals have resulted in a few mishaps from improper use. Adequate safety-in-use evaluations of any of these chemicals persisting into our air, drinking water, and foodstuffs are not simple matters, and they incorporate the considered judgments of many individuals highly trained in a variety of complex biological, chemical, food technological, medical, pharmacological, and toxicological disciplines.

It is hoped that *Reviews of Environmental Contamination and Toxicology* will continue to serve as an integrating factor both in focusing attention upon those matters requiring further study and in collating for variously trained readers present knowledge in specific important areas involved with chemical contaminants in the total environment. This and previous volumes of "Reviews" illustrate these objectives. Because manuscripts are published in the order in which they are received in final form, it may seem that some important aspects of analytical chemistry, bioaccumulation, biochemistry, human and animal medicine, legislation, pharmacology, physiology, regulation, and toxicology are being neglected. To the contrary, these apparent omissions are recognized, and some pertinent manuscripts are in preparation. However, the field is so large and the interests in it are so varied that the editor and the Editorial Board earnestly solicit suggestions of topics and authors to help make this international book-series even more useful and informative.

Reviews of Environmental Contamination and Toxicology attempts to provide concise, critical reviews of timely advances, philosophy, and significant areas of accomplished or needed endeavor in the total field of foreign chemicals in any segment of the environment, as well as toxicological implications. These reviews are either general or specific, but properly they may lie in the domains of analytical chemistry and its methodology, biochemistry, human and animal medicine, legislation, pharmacology, physiology, regulation, and toxicology. Certain affairs in the realm of food technology concerned specifically with pesticide and other food-additive problems are also appropriate subject matter.

The justification for the preparation of any review for this book-series is that it deals with some aspect of the many real problems arising from the presence of any "foreign" chemicals in our surroundings. Thus, manuscripts may encompass those matters in any country. Added plant or animal pest-control chemicals or their metabolites that may persist into food and animal feeds are within this scope. The so-called food additives (substances deliberately added to foods for flavor, odor, appearance, and preservation, as well as those inadvertently added during manufacture, packing, distribution, and storage) are also considered considered suitable review material. In addition, chemicals contaminant in any manner to air, water, soil, or plant or animal life are within this purview and these objectives.

Manuscripts are normally contributed by invitation but suggested topics are welcome. Preliminary communication with the editor is recommended before volunteered reviews are submitted in manuscript form.

College of Agriculture   G.W.W.
University of Arizona
Tucson, Arizona

Table of Contents

Microbial Metabolism of Pesticides and Structurally Related Compounds

Ian C. MacRae*

Contents

---

*Department of Microbiology, University of Queensland, St. Lucia, Brisbane, Australia, 4067.

© 1989 Springer-Verlag New York Inc.
Reviews of Environmental Contamination and Toxicology, Vol. 109.

I. Introduction

Synthetic organic pesticides have been in widespread use for more than 40 years, and during that period, their use has contributed greatly to increased world-wide food production and improved human and animal health. However, these successes have not been without their side effects such as toxicities to non-target species, including humans, and the production of persistent residues in soil and water.

The importance of microorganisms in the degradation and detoxication of pesticides was soon established, and biodegradation, chiefly by microorganisms, was recognized as a major means of destroying these chemicals in soil and water. This review brings together published literature in the period 1981 to 1987 on the metabolism of pesticides and structurally related compounds in soil and water. Reviews on pesticide degradation in general, published during this period, include those of Alexander (1981, 1985), Hutzinger and Veerkamp (1981), Matsumura (1982), Sethunathan et al. (1982), Slater and Bull (1982), Dagley (1983), Ghisalba (1983), Leisinger (1983), Motosugi and Soda (1983), and Barik (1984).

During this period much has happened in research on pesticide metabolism. New methods and new approaches to old problems have been devised, and in some cases their use has led to the confirmation and expansion of earlier knowledge. Genetic engineering has become a promising tool for those concerned with detoxication of soil and water and the disposal of toxic wastes, and soil microbes appear to have learned new tricks, such that enhanced degradation of pesticides has become a serious problem in some areas.

II. Methodology

The traditional approach to microbial biodegradation studies often involved batch enrichment culture, followed by the isolation of pure cultures of microorganisms able to grow on the pesticide as a sole source of carbon and energy. Although this approach has yielded valuable information on the role of microbes in the degradation of pesticides it does have serious limitations and by no means reveals the full involvement of microbes in these processes.

Major problems associated with conventional enrichment culture in biodegradation studies include: (1) isolates obtained may or may not be significant in the degradation of the pesticide in nutrient-poor natural environments because the enrichment pressure in the batch culture at relatively high pesticide concentration is related to the maximum specific growth rates of the various microbes in the inoculum; (2) some pesticides may undergo partial or extensive degradation by cometabolism and therefore the active microbial population will require a growth substrate; (3) the pesticide degradation may require the complex interplay of several members of a microbial consortium which often leads to frustrated attempts by the microbiologist to isolate pure cultures able to degrade the pesticide; (4) at the relatively high levels of pesticide usually employed for enrichment culture, the pesticide and/or its transformation products may be toxic to microbes in the inoculum; and (5) environmental interfaces, alternating changes in the aeration status or redox potential of the environment and/or gradients may be needed to promote biodegradation. These problems have been discussed at length by Bull (1980), Harder (1981), and Cook et al. (1983).

Continuous culture systems offer a number of important advantages over batch cultures that are useful in biodegradation studies. These include the ability to study the effects of (1) changing substrate concentration, (2) mixed substrates, (3) biodegradation by mixed cultures, (4) strong selective pressures on the microbial community in the culture over long periods, and (5) alternating aerobic and anaerobic conditions (Harder 1981).

Although quite a few groups have used continuous culture in biodegradation studies, one of the success stories was that of Kellogg et al. (1981). Beginning with low 2,4,5-T concentrations in the presence of other substrates, such as toluate, salicylate, and chlorobenzoate, aimed at maintaining high plasmid levels in the chemostat enrichment, they developed after 8 to 10 mon of continuous culture a microbial population that could degrade 2,4,5-T. The process was termed "plasmid-assisted molecular breeding." Subsequent isolations from the chemostat culture (Kilbane et al. 1982) yielded a pure culture, designated *Pseudomonas cepacia*, that could use 2,4,5-T as a sole carbon source for growth, something that many workers had unsuccessfully attempted using conventional batch enrichment.

While pure culture studies have been numerous over this review period, emphasis has been placed on biodegradation employing mixed natural microbial communities as well as artificial mixed cultures. Some effort has assessed laboratory methods used for examining the biodegradability of organic compounds and predicting rates of biodegradation in natural environments. Also, emphasis has been placed on the use of microcosms as laboratory models to gain an understanding of biodegradation in the field.

To overcome problems associated with extrapolating laboratory results to natural environments, Harvey (1983) devised a method for evaluating soil degradation of ^{14}C-labeled pesticides under field conditions. Laskowski et al. (1983) have discussed laboratory evaluation of soil degradation. Recognizing the complexity of biological processes involved in biodegradation of organic compounds in natural environments, Liu et al. (1981a) modified a cyclone fermenter to study the biodegradation of pesticides under aerobic or anaerobic conditions and with and without cometabolites.

Spain et al. (1984) compared the biodegradation of *p*-nitrophenol in various laboratory systems, including shake flasks, ecocores, and microcosms, with biodegradation of the phenol in a freshwater pond. They found that all systems gave results in good agreement with field data in terms of the major trends in the biodegradation of *p*-nitrophenol. However, there were considerable differences in rates among the systems and the authors identified aeration and mixing rates as important factors leading to differences in degradation rates.

Shelton and Tiedje (1984a) have examined the problem of measuring the anaerobic biodegradation potential of more than 100 organic chemicals. They found that the use of a pressure transducer to measure the net increase in gas pressure which developed in sealed bottles during CO_2 and CH_4 production from the test chemical gave quick, inexpensive, and accurate results. Other systems described for biodegradation studies included an ultrafiltration cell device (Bengtsson et al. 1986) and an ecocore system (Houx and Dekker 1987).

The use of radiolabeled chemicals, particularly ^{14}C-labeled substances, in biodegradation is not new but some methods have been refined or modified in the last 8 yr. For example, Pfaender and Bartholomew (1982) obtained rates of biodegradation of pollutants in various aquatic environments using a modified heterotrophic uptake technique, measuring substrate incorporation into biomass as well as the amount of substrate mineralized.

Modifications of a double-vial radio respirometric technique, which is a convenient way to assess the mineralization of organic chemicals, have been described (McKinley et al. 1983; Coveney and Wetzel 1984; Deeley et al. 1985). Somerville et al. (1985) used nonpolar and volatile substances to give information on most-probable-number (MPN) estimates of degrader populations.

In the past biodegradation of pesticides and other pollutants was assessed in laboratory evaluations using chemical concentrations generally far in excess of those normally encountered in natural environments. Extrapolation of results obtained in this way may not reflect the true situation. Because of the sensitivity of detection, the use of radiolabeling, usually with ^{14}C in the carbon skeleton, has enabled research into transformations of trace amounts, yielding useful information on the mineralization and/or cometabolism at concentrations commonly found in natural environments. Other factors, such as experimental design, must be considered in the evaluation of biodegradation potential of a chemical in natural ecosystems. Fannin et al. (1981) stressed the importance of statistical design and suggested the use of a fractional factorial design.

III. Microbial Metabolism of Specific Groups of Pesticides

During this review period, research into the microbial metabolism of a diverse range of pesticides has been conducted. Also, the metabolism of pesticides, their degradation products, and structurally similar compounds has been studied using many different systems, including pure microbial cultures, defined mixed populations, naturally occurring microbial consortia, wastewater, sludge, soil and water microcosms, immobilized microbial cells, and microbial enzymes. These studies are summarized in Table 1.

A. Halogenated Hydrocarbons and Related Substances

Halogenated hydrocarbon pesticides, commonly referred to as organochlorines, and their oxy-derivatives have long been known to cause environmental problems because of their persistence, toxicities, and carcinogenic properties. Therefore, it is surprising that many of these are still marketed for agricultural and domestic purposes. For example, use of BHC (active isomer, γ-1,2,3,4,5,6-hexachlorocyclohexane) has been banned for about 10 yr in a number of countries, but extensive use of this insecticide was made in sugarcane production in Australia and India until recently. Soap formulations containing lindane for use in controlling head lice infestations in children are still available.

Although there has been a trend in some countries to phase out halogenated hydrocarbon insecticides and replace them with less toxic and less recalcitrant alternatives, significant soil residues are still detectable. These residues not only pose environmental problems but might also affect human and animal health as well as the marketability of products from contaminated areas. For example, soil that has residues of BHC built up through its use in sugar cane production might, if put to alternate use in vegetable production or pastures or if used for aquaculture, lead to levels in these products or in the tissues of grazing animals that are unacceptable. Therefore, although few over the past 9 years, studies on the microbial metabolism of these compounds are still relevant to management of the environment. The microbial metabolism of halogenated insecticides has been reviewed by Lal and Saxena (1982) and Barik (1984).

1. Chlordane. As a protection agent for wood against termites, chlordane has been very successful, and it does have the long persistence and toxicity problems common in other chlorinated pesticides. However, it is susceptible to biodegradation by a soil bacterium, *Nocardiopsis* sp. (Beeman and Matsumura 1981). Pure, growing cultures of this actinomycete bring about extensive degradation of both *cis*- and *trans*-chlordane. Beeman and Matsumura (1981) characterized chemically at least eight products of chlordane metabolism including dichlorochlordene, oxychlordene, heptachlor-endo-epoxide, chlordene chlorohydrin, and 3-hydroxy-*trans*-chlordane.

Table 1. Microbial metabolism of pesticides and their degradation products.

Functional Group	Pesticide or Degradation Product*	Active Microbial Population	References
Halogenated hydrocarbons	Chlordane (I)	*Nocardiopsis* sp.	Beeman and Matsumura (1981)
	DDT (I)	*Alcaligenes* sp.	Subba-Rao and Alexander (1985)
		Pseudomonas sp.	
		Penicillium sp.	
		Aspergillus conicus	
		Aspergillus niger	
		Penicillium brefeldianum	
		Pseudomonas aeruginosa	Golovleva and Skryabin (1981)
		Soil microbes	Fuhremann and Lichtenstein (1980)
		Phanerochaete chrysosporium	Bumpus and Aust (1987)
		Phellinus weirii	
		Pleurotus ostreatus	
	Hexachlorocyclohexane (I)	*Clostridium rectum*	Ohisa et al. (1980 and 1982)
			Kurihara et al. (1981)
		Soil microbes	Fuhremann and Lichtenstein (1980)
			MacRae et al. (1984)
		Mixed anaerobic population	Maule et al. (1987)
	Mirex (I)	*Bacillus sphaericus*	Aslanzadeh and Hedrick (1985)
		Streptomyces albus	
Halogenated hydrocarbons -oxy derivatives	Dieldrin (I)	*Bacillus* spp.	Singh (1981)
		Micrococcus spp.	
		Yeasts	
		Clostridium bifermentans	Maule et al. (1987)
		Clostridium glycolium	
		Clostridium sp.	

	Methoxychlor (I)	*Klebsiella pneumoniae*	Baarschers et al. (1982)
		Soil microbes	Fogel et al. (1982)
	Halo-aromatics	*Pseudomonas cepacia*	Vandenbergh et al. (1981)
		Alcaligenes sp.	De Bont et al. (1986)
		Pseudomonas sp.	Spain and Nishino (1987)
		Rhodosporidium sp.	Corbett and Corbett (1981)
		Freshwater microbes	Bartholomew and Pfaender (1983)
		Estuarine water microbes	
		Marine water microbes	
		Denitrifying bacteria	Bouwer and McCarty (1982)
Hydrocarbons	Methylnaphthalenes	*Cunninghamella elegans*	Cerniglia et al. (1984)
		Soil column microbes	Hutchins et al. (1984b)
	Anthracene	Marine sediment microbes	Bauer and Capone (1985)
	Naphthalene	Marine sediment microbes	
		Groundwater microbes	Ehrlich et al. (1982)
Phosphates	Chlorfenvinfos (I)	Soil microbes	Miles et al. (1984)
	Dichlorvos (I)	Sewage microbes	Lieberman and Alexander (1983)
	Fenitrooxon (I)	*Trichoderma viride*	Baarschers and Heitland (1986)
Phosphorothioates	Chlorpyrifos (I)	Soil microbes	Getzin (1981)
			Leoni et al. (1981)
			Miles et al. (1984)
		Estuarine water microbes	Schimmel et al. (1983)
	Coumaphos (I)	*Flavobacterium* sp.	Kearney et al. (1986)
	Diazinon (I)	*Flavobacterium* sp.	Adhya et al. (1981b)
			Forrest et al. (1981)
		Pseudomonas sp.	Adhya et al. (1981b)
		Pseudomonas aeruginosa	Merritt et al. (1981)

Table 1. (*Continued*)

Functional Group	Pesticide or Degradation Product*	Active Microbial Population	References
Phosphorothioates (*continued*)	Fenitrothion (I)	*Flavobacterium* sp.	Adhya et al. (1981b)
		Trichoderma viride	Baarschers and Heitland (1986)
		Flooded soil microbes	Adhya et al. (1981a)
		Estuarine water microbes	Weinberger et al. (1982b)
		Lake water microbes	Weinberger et al. (1982b)
		Distilled water microcosm	
		Soil microbes	MacRae (1986b)
	Fensulfothion (I)	*Alcaligenes* sp.	Sheela and Pai (1983)
		Hafnia sp.	MacRae and Cameron (1985)
		Klebsiella pneumoniae	Timms and MacRae (1982 and 1983)
		Pseudomonas alcaligenes	Sheela and Pai (1983)
		Mixed bacterial culture	Miles and Moy (1982)
	Fenthion ethyl (I)	*Pseudomonas aeruginosa*	Merritt et al. (1981)
	Methyl parathion (I)	*Flavobacterium* sp.	Adhya et al. (1981b)
			Lewis et al. (1985)
		Pseudomonas sp.	Adhya et al. (1981b)
		Freshwater microbes	VanVeld and Spain (1983)
		Estuarine water microbes	Schimmel et al. (1983)
		Flooded soil microbes	Adhya et al. (1981a and c)
		Soil microbes	Ou et al. (1983 and 1985)
		Flavobacterium aquatile	Lewis and Holm (1981)
		Staphylococcus saprophyticus	
		Aufwuchs population	
	Parathion (I)	*Arthrobacter* sp.	Nelson (1982)
		Bacillus sp.	

		Flavobacterium sp.	Adhya et al. (1981b)
			Forrest et al. (1981)
			Mulbry et al. (1986)
		Pseudomonas sp.	Adhya et al. (1981b)
		Pseudomonas diminuta	Serdar et al. (1982)
		Rice rhizosphere microbes	Reddy and Sethunathan (1983)
		Flooded soil microbes	Reddy and Sethunathan (1985)
			Adhya et al. (1981a)
		Soil microbes	Fuhremann and Lichenstein (1980)
			Lichtenstein et al. (1982 and 1983)
			Nelson et al. (1982)
		Activated sludge	Laplanche et al. (1981)
Phosphorodithioates	Fonofos (I)	Soil microbes	Fuhremann and Lichtenstein (1980)
			Huckins et al. (1986)
	Malathion (A)	Freshwater microbes	Paris et al. (1981)
	Phorate (I)	Soil microbes	Fuhremann and Lichtenstein (1980)
			Chapman et al. (1982b)
	Terbufos (I)	Soil microbes	Chapman et al. (1982a)
Phosphoramidates	Glyphosate (H)	Arthrobacter sp.	Pipke et al. (1987)
		Flavobacterium sp.	Balthazor and Hallas (1986)
		Pseudomonas sp.	Moore et al. (1983)
			Shinabarger et al. (1984)
		Activated sludge microbes	Balthazor and Hallas (1986)
		Soil microbes	
Phosphoramidothioates	Isofenphos (I)	Soil microbes	Chapman et al. (1986c)
Carbamates	Carbaryl (I)	Bacillus sp.	Rajagopal et al. (1984b)
		Pseudomonas sp.	Larkin and Day (1985 and 1986)
		Rhodococcus sp.	Larkin and Day (1986)

Table 1. (*Continued*)

Functional Group	Pesticide or Degradation Product*	Active Microbial Population	References
Carbamates *(continued)*	Carbaryl (I) *(continued)*	Anaerobes	Kiene and Capone (1986)
		Flooded soil microbes	Rajagopal et al. (1983)
			Rajagopal and Sethunathan (1984)
		Soil enrichment	Rajagopal et al. (1984b)
	Carbofuran (I)	*Arthrobacter* sp.	Rajagopal et al. (1984b)
		Azospirillum lipoferum	Venkateswarlu and Sethunathan (1984)
		Bacillus sp.	Rajagopal et al. (1984b)
		Nocardia sp.	Venkateswarlu and Sethunathan (1985)
		Pseudomonas cepacia	Venkateswarlu and Sethunathan (1985)
		Pseudomonas sp.	Felsot et al. (1981)
		Streptomyces sp.	Venkateswarlu and Sethunathan (1984)
		Soil microbes	Fuhremann and Lichtenstein (1980)
			Miles et al. (1981)
			Harris et al. (1984)
			Chapman et al. (1986a and b)
		Flooded soil microbes	Rajagopal and Sethunathan (1984)
			Rajagopal et al. (1986)
	Methomyl (I)	Methanogenic microbes	Kiene and Capone (1986)
	CIPC (H)	*Anacystis nidulans*	Wright and Maule (1982)
		Pseudomonas cepacia	Vega et al. (1985)
		Freshwater microbes	Paris et al. (1981)
	IPC (H)	Sewage microbes	Wang et al. (1984)
		Lake water microbes	Hoover et al. (1986)

Promecarb (I)	*Aeromonas liquefaciens*	Knowles and Benezet (1981)
	Bacillus cereus	
	Flavobacterium sp.	
	Proteus vulgaris	
	Pseudomonas putida	
Phenmedipham (H)	*Bacillus cereus*	Knowles and Benezet (1981)
	Bacillus subtilis	
	Flavobacterium sp.	
	Proteus vulgaris	
	Pseudomonas putida	
	Aspergillus versicolor	
	Torula rosea	
Desmedipham (H)	*Enterobacter aerogenes*	
	Aeromonas liquefaciens	
	Bacillus cereus	
	Bacillus megaterium	
	Bacillus subtilis	
	Flavobacterium sp.	
	Proteus vulgaris	
	Pseudomonas putida	
	Aspergillus versicolor	
	Penicillium cyclopium	
	Torula rosea	

Table 1. (*Continued*)

Functional Group	Pesticide or Degradation Product*	Active Microbial Population	References
Desmedipham (H) (*continued*)	Cycloate (H)	Sewage microbes	Novick and Alexander (1985)
		Lakewater microbes	
Thiocarbamates	Diallate (H)	Soil microbes	Anderson (1984)
	Triallate (H)	Soil microbes	
	EPTC (H)	*Alcaligenes* sp.	Lee (1984)
		Arthrobacter sp.	Tam et al. (1987)
		Bacillus sp.	Lee (1984)
		Epicoccum sp.	
		Micrococcus sp.	
		Pseudomonas sp.	
		Chaetomium globosum	
		Diheterospora sp.	
		Fusarium oxysporum	
		Fusarium solani	
		Paecilomyces sp.	
		Penicillium sp.	
		Trichoderma sp.	
		Verticillium sp.	
Triazines	Ametryne (H)	Fluorescent pseudomonad	Cook and Hutter (1982)
	Atrazine (H)	*Nocardia* sp.	Giardina et al. (1982)
		Pseudomonas sp.	Behki and Khan (1986)
		Facultative anaerobe	Jessee et al. (1983)
	Metamitron (H)	*Arthrobacter* sp.	Engelhardt et al. (1982)
	Prometryn (H)	Fluorescent pseudomonad	Cook and Hutter (1982)

Triazines (H)	*Klebsiella pneumoniae*	Cook and Hutter (1981a)
	Pseudomonas sp.	Cook and Hutter (1981a)
		Grossenbacher et al. (1984)
Phenoxys	2,4-D (H)	
	Alcaligenes sp.	Amy et al. (1985)
	Alcaligenes eutrophus	MacRae (1985)
	Anaerobic sludge microbes	Mikesell and Boyd (1985)
		Gibson and Suflita (1986)
	Sewage microbes	Subba-Rao et al. (1982)
		Hoover et al. (1986)
	Freshwater microbes	Subba-Rao et al. (1982)
		DeLaune and Salinas (1985)
	Lakewater microbes	Hoover et al. (1986)
		Wiggins et al. (1987)
	Soil microbes	Duah-Yentumi and Kuwatsuka (1982)
		Fournier (1980)
		Fournier et al. (1981)
		Soulas and Fournier (1981)
		Kunc and Rybarova (1983)
		Parker and Doxtader (1983)
		Soulas et al. (1983 and 1984)
		Stott et al. (1983)
		Ogram et al. (1985)
		Ou (1984)
		Smith (1985a)
	Mixed bacterial culture	Kilpi (1980)
	Water and sediment ecocore microbes	Spain and Van Veld (1983)

Table 1. (*Continued*)

Functional Group	Pesticide or Degradation Product*	Active Microbial Population	References
Phenoxys (*continued*)	2,4-D ethyl sulfate (H)	*Pseudomonas putida*	Lillis et al. (1983)
	2,4-D Butoxy ethyl ester (H)	Freshwater microbes	Paris et al. (1981)
		Microbial community	Lewis et al. (1983 and 1984)
	2,4,5-T (H)	*Pseudomonas cepacia*	Chatterjee et al. (1982)
			Kilbane et al. (1982)
			Karns et al. (1983b)
			Kilbane et al. (1983)
		Anaerobic consortium	Suflita et al. (1984)
		Anaerobic sludge microbes	Mikesell and Boyd (1985)
		Soil microbes	Kilbane et al. (1983)
			McCall et al. (1981)
	Mecoprop (H)	Consortium of *Pseudomonas* spp., *Alcaligenes* sp., *Flavobacterium* sp., *Acinetobacter calcoaceticus*	Lappin et al. (1985)
		Soil microbes	Smith and Hayden (1981)
			Smith (1985b)
		Mixed bacterial culture	Kilpi (1980)
	MCPA (H)	Consortium of *Pseudomonas* spp. *Alcaligenes* sp., *Flavobacterium* sp. *Acinetobacter calcoaceticus*	Lappin et al. (1985)
		Soil microbes	Duah-Yentumi and Kuwatsuka (1982)
			Smith and Hayden (1981)
			Soulas et al. (1983)
		Mixed bacterial culture	Kilpi (1980)

	MCPB (H)	Soil microbes	Smith and Hayden (1981)
Phenols	Phenol	*Pseudomonas acidovorans*	Schmidt and Alexander (1985)
			Schmidt et al. (1985)
		Pseudomonas putida	Paris et al. (1982)
			Molin and Nilsson (1985)
		Streptomyces setonii	Antai and Crawford (1983)
		Methanogenic bacteria	Young and Rivera (1985)
		Activated sludge microbes	Beltrame et al. (1984)
			Hannah et al. (1986)
		Sludge microbes	Horowitz et al. (1982)
			Hannah et al. (1986)
		Anaerobic sludge microbes	Boyd et al. (1983)
		Sewage microbes	Subba-Rao et al. (1982)
			Rubin et al. (1982)
			Deeley et al. (1985)
			Simkins et al. (1986)
		Landfill leachate microbes	Deeley et al. (1985)
		Lakewater microbes	Rubin et al. (1982)
			Rubin and Alexander (1983)
			Jones and Alexander (1986)
		Freshwater microbes	Rubin et al. (1982)
			Subba-Rao et al. (1982)
			Chesney et al. (1985)
			Rubin and Schmidt (1985)

Table 1. (*Continued*)

Functional Group	Pesticide or Degradation Product*	Active Microbial Population	References
Phenols (*continued*)	Phenol (*continued*)	Sediment microbes	Horowitz et al. (1982)
		Anaerobic groundwater microbes	Ehrlich et al. (1982)
		Biofilm continuous culture	Molin and Nilsson (1985)
		Mixed aquatic microcosm	Shrimp and Pfaender (1987)
		Immobilized methanogens	Dwyer et al. (1986)
	Cresol	Freshwater microbes	Pfaender and Bartholomew (1982)
			Van Veld and Spain (1983)
		Estuarine water microbes	Bartholomew and Pfaender (1983)
		Heterogeneous culture	Masunaga et al. (1986)
		Water and sediment ecocore microbes	Spain and Van Veld (1983)
		Bacterial consortium	Suflita et al. (1987)
		Anaerobic aquifer microbes	Smolenski and Suflita (1987)
Substituted Phenols	Pentachlorophenol (I,F,H)	*Arthrobacter* sp.	Stanlake and Finn (1982)
			Neilson et al. (1983)
			Edgehill and Finn (1983)
		Flavobacterium sp.	Saber and Crawford (1985)
			Brown et al. (1986)
			Steiert et al. (1987)
		Pseudomonas cepacia	Karns et al. (1983b)
		Pseudomonas sp.	Trevors (1982)
		Sludge microbes	Hannah et al. (1986)
		Anaerobic sludge microbes	Mikesell and Boyd (1985 and 1986)
			Godsy et al. (1986)
		Sewage sludge	Guthrie et al. (1984)
			Moos et al. (1983)

	Freshwater microbes	Pignatello et al. (1983)
		Pignatello et al. (1986)
	Flooded soil microbes	Weiss et al. (1982)
	Epilithic consortium	Brown et al. (1986)
	Mixed bacteria and protozoa	Klecka and Maier (1985)
	Microbial consortium	Salkinoja-Salonen et al. (1983)
		Apajalathi and Salkinoja-Salonen (1984)
Halophenols	*Acinetobacter* sp.	Allard et al. (1987)
	Alcaligenes sp.	Schmidt et al. (1983)
		Schwien and Schmidt (1982)
	Arthrobacter sp.	Neilson et al. (1983)
	Flavobacterium sp.	Steiert et al. (1987)
		Schwien and Schmidt (1982)
	Pseudomonas sp.	Schmidt et al. (1983)
		Goldstein et al. (1985)
		Spain and Nishino (1987)
	Pseudomonas cepacia	Karns et al. (1983a and b)
		Kilbane et al. (1982)
	Rhodococcus sp.	Allard et al. (1987)
	Soil microbes	Sattar (1981)
	Activated sludge microbes	Hannah et al. (1986)
	Anaerobic sludge	Boyd and Shelton (1984)
	Defined mixed flora	Schmidt et al. (1983)
Halocatechols	Soil microbes	Sattar (1981)
		Stott et al. (1983)
		Cheng et al. (1983)
Nitrophenols	*Pseudomonas* sp.	Goldstein et al. (1985)
	Pseudomonas putida	Zeyer et al. (1986a)
	Sewage microbes	Rubin et al. (1982)
		Hoover et al. (1986)

Table 1. (*Continued*)

Functional Group	Pesticide or Degradation Product*	Active Microbial Population	References
Substituted phenols (*continued*)	Nitrophenols (*continued*)	Lakewater microbes	Rubin et al. (1982)
			Hoover et al. (1986)
		Estuarine water microbes	Van Veld and Spain (1983)
		Activated sludge microbes	Kool (1984)
		Water and sediment ecocore microbes	Spain and Van Veld (1983)
		Flavobacterium sp.	Wiggins et al. (1987)
		Bacteria on activated carbon	De Laat et al. (1985)
Anilines	Aniline	*Escherichia coli*	Lammerding et al. (1982)
		Pseudomonas sp.	Kaminski et al. (1983)
			Anson and McKinnon (1984)
			Schmidt and Alexander (1985)
	Aniline	*Rhodococcus* sp.	Kaminski et al. (1983)
		Sewage microbes	Hoover et al. (1986)
		Lakewater microbes	Hoover et al. (1986)
		Sediment microbes	Pillai et al. (1982)
		Pond water	Lyons et al. (1984)
Substituted Anilines	Haloanilines	*Escherichia coli*	Bunce et al. (1983)
		Moraxella sp.	Zeyer et al. (1985)
		Pseudomonas sp.	Zeyer and Kearney (1982a and b)
			Kaminski et al. (1983)
		Pseudomonas putida	You and Bartha (1982a and b)
		Rhodococcus sp.	Kaminski et al. (1983)
		Trametes versicolor	Hoff et al. (1985)
		Rhizoctonia praticola	

Class	Pesticide	Organism	Reference
		Freshwater microbes	Hwang et al. (1985)
		Mixed bacteria in water	Paris and Wolfe (1987)
	Nitroaniline	*Pseudomonas* sp.	Zeyer and Kearney (1983a)
Amides	Alachlor (H)	Sewage microbes	Novick and Alexander (1985)
		Lakewater microbes	
		Anaerobic stream sediment	Bollag et al. (1986)
	Propachlor (H)	Sewage microbes	Novick and Alexander (1985)
		Lakewater microbes	
		Soil microbes	Novick et al. (1986)
	Propanil (H)	*Anabaena cylindrica*	Wright and Maule (1982)
		Anacystis nidulans	
		Chlamydomonas reinhardii	
		Chlorella vulgaris	
		Gloeocapsa alpicola	
		Tolypothrix tenuis	
		Pseudomonas sp.	Zeyer and Kearney (1982b)
		Sediment microbes	Stepp et al. (1985)
		Soil anaerobes	Pettigrew et al. (1985)
Ureas	Chlorbromuron (H)	Sediment microbes	Stepp et al. (1985)
	Diuron (H)	Sediment microbes	Stepp et al. (1985)
			Attaway et al. (1982)
	Linuron (H)	Sediment microbes	Stepp et al. (1985)
	Diflubenzuron (I)	Soil microbes	Nimmo et al. (1984 and 1986)
			Chapman et al. (1985)

Table 1. (*Continued*)

Functional Group	Pesticide or Degradation Product*	Active Microbial Population	References
Metallo & Metallo Organics	Bis (tributyltin) oxide (F)	*Alcaligenes faecalis* *Pseudomonas aeruginosa* *Chaetomium globosum* *Coniophora puteana* *Trametes versicolor*	Barug (1981)
	Chromated copper arsenate	*Candida humicola*	Cullen et al. (1984)
	Mancozeb (F)	*Bacillus* sp.	Doneche et al. (1983)
Pyrethroids	Cypermethrin (I)	Soil microbes	Chapman et al. (1981)
	Decamethrin (I)	Soil microbes	
	Fluvalinate (I)	Flooded soil microbes	Staiger and Quistad (1983)
	Fenpropanate (I)	Soil microbes	Chapman et al. (1981)
	Fenvalerate (I)	Soil microbes	Chapman et al. (1981) Smith and Willis (1985)
		Estuarine water microbes	Schimmel et al. (1983)
	Permethrin (I)	Soil microbes	Chapman et al. (1981) Doyle et al. (1981)
		Estuarine water microbes	Schimmel et al. (1983)
Sulfides	p-chlorophenylmethyl sulfide	Soil microbes	Guenzi and Beard (1981)
Imides	Benzazimide	*Pseudomonas* sp.	Engelhardt and Wallnofer (1983)

Triazolines	Triadimefon (F)	*Botrytis cinerea* *Cladosporium cucumerinum* *Coriolus versicolor* *Fusarium culmorum*	Deas et al. (1984a and b)
Ketones	Warfarin (R)	*Arthrobacter* sp. *Nocardia corallina*	Davis and Rizzo (1982)
Thiuram	TMTD (F)	*Pseudomonas aeruginosa* Soil microbes	Shirkot and Gupta (1985)
Botanicals	1',2'-Dihydrorotenone (I)	*Streptomyces griseus*	Sariaslani and Rosazza (1985)
Formamidines	Chlordimeform (M)	*Oscillatoria* sp. *Chlorella* sp.	Benezet and Knowles (1981)
Nitriles	Bromoxynil (H)	*Klebsiella pneumoniae*	McBride et al. (1986)
	Dichlorbenil (H)	Soil microbes	Chowdhury et al. (1981)
Nitro-	Profluralin (H)	*Arthrobacter simplex* *Cellulomonas flavigenum* *Microbacterium flavum*	Stralka and Camper (1981)
	Trifluralin (H)	*Candida* sp.	Zeyer and Kearney (1983b)
	Pentachloronitrobenzene (F)	*Tetrahymena thermophila*	Murphy et al. (1982)
	Nitrobenzene	Sewage microbes	Hallas and Alexander (1983)
Pyridiliums	Paraquat (H)	*Lipomyces starkeyi*	Carr et al. (1985)

Table 1. (*Continued*)

Functional Group	Pesticide or Degradation Product*	Active Microbial Population	References
Toluidides	Metolachlor (H)	*Bacillus circulans*	Saxena et al. (1987)
		Bacillus megaterium	
		Fusarium sp.	
		Mucor racemosus	
		Actinomycete	Krause et al. (1985)
Carboxylic Acids	Halobenzoic Acids	*Alcaligenes denitrificans*	van den Tweel et al. (1987)
		Arthrobacter sp.	Marks et al. (1984)
		Pseudomonas sp.	Veerkamp et al. (1983)
		Pseudomonas fluorescens	Johnson and Williams (1982)
		Chlamydomonas sp.	Jacobson and Alexander (1981)
		Lake sediment microbes	Horowitz et al. (1983)
			Suflita et al. (1983)
		Pseudomonas alcaligenes	Focht and Shelton (1987)
	Fenac (H)	Soil microbes	Rosenberg (1984)
	Endothal (H)	Freshwater microbes	Reinert et al. (1986)

*A: acaricide; F: fungicide; H: herbicide; I: insecticide; M: miticide; R: rodenticide.

2. DDT.   Degradation studies using a soil-oat plant system, by Fuhremann and Lichtenstein (1980), revealed that ^{14}C-labeled DDT was only slowly degraded to *p,p'*-DDE, TDE, and dicofol. However, complete degradation of DDT by a pure culture of *Pseudomonas aeruginosa* strain 640X was possible if a complex sequence of alternating cosubstrates, gradients, and aeration conditions were applied to the culture (Golovleva and Skryabin 1981). Apart from the first step in the degradation of DDT by this bacterium, a reductive dechlorination to remove a chlorine atom from the trichloroethane bridge, cosubstrates were required for all other steps until the formation of benzylhydrol. The best cosubstrate proved to be hexadecane. The authors felt that while complete degradation of DDT was possible in the laboratory, it would be impossible to provide the complex sequence of conditions to bring about complete DDT degradation in natural environments. This outlook was not shared by Subba-Rao and Alexander (1985), who felt that extensive degradation of DDT in culture and in nature was likely.

Recently, Bumpus and Aust (1987) have demonstrated extensive degradation and mineralization of DDT in cultures of the white rot fungus, *Phanerochaete chrysosporium*. They detected the formation of various metabolites in the cultures and these too were metabolized. Those metabolites that were characterized by gas chromatography-mass spectrometry (GC-MS) included 1,1-dichloro-2,2-bis(4-chlorophenyl)ethane (DDD), 2,2,2-trichloro-1,1-bis(4-chlorophenyl) ethanol (dicofol), 2,2-dichloro-1,1-bis(4-chlorophenyl)ethanol, and 4,4'-dichlorobenzophenone. The pathway of degradation of DDT by this fungus is different from the ones published for bacteria. Bumpus and Aust (1987) found that some other white rot fungi such as *Pleurotus ostreatus*, *Phellinus weirii*, and *Polyporus versicolor* could also mineralize DDT.

Possible alternatives to DDT for use in pest control with lower toxicity and better biodegradability may be those compounds consisting of a DDT isostere molecule linked to a pyrethrin structure (Ralph 1986).

3. Hexachlorocyclohexane.   The metabolism of lindane (γ-isomer of 1,2,3,4,5, 6-hexachlorocyclohexane) including microbial metabolism of lindane has been reviewed by Macholz and Kujawa (1985). Studies on the metabolism of γ-HCH and related compounds by cell-free extracts of the anaerobe *Clostridium rectum* (Ohisa et al. 1980 and Kurihara et al. 1981) and whole cells of the same bacterium (Ohisa et al. 1982) showed these compounds serve as artificial electron acceptors in the Stickland reaction leading to ATP production.

Fuhremann and Lichtenstein (1980) obtained chromatographic evidence that led them to suspect the formation of γ-pentachlorocyclohexane during the degradation of ^{14}C-lindane in soil. The addition of a mixture of amino acids, yeast extract, or inositol to anaerobically incubated suspensions of a volcanic ash soil increased disappearance of the α-, β-, and γ-isomers of hexachlorocyclohexane (MacRae et al. 1984).

4. Mirex.   By the conventional enrichment technique, Aslanzadeh and Hedrick (1985) isolated two bacteria in pure culture, *Bacillus sphaericus* and *Streptomyces albus*, that could mineralize mirex.

5. Oxy- derivatives.   Among the oxy-derivatives of the halogenated insecticides, methoxychlor is a replacement for DDT. Fogel et al. (1982) found that sequential anaerobic and aerobic soil conditions favored the mineralization of methoxychlor ring carbon to CO_2 and suspected that cometabolism was important in the degradative process. The initial step in degradation, dechlorination, required anaerobic conditions. They reasoned that since many soils and sediments undergo periodic changes in redox potential, the sequential conditions needed for methoxychlor mineralization may thus be provided leading to degradation in natural environments. Pure culture studies by Baarschers et al. (1982) with *Klebsiella pneumoniae* showed that the bacterium had the ability to dechlorinate reductively methoxychlor and its dehydroxy derivative [2,2-bis (*p*-hydroxyphenyl)-1,1,1-trichloroethane] to the corresponding dichloroderivatives.

Microflora on the surface of blowflies yielded cultures of *Bacillus* spp., *Micrococcus* spp., and yeasts that were able to metabolize ^{14}C-dieldrin when added to nutrient broth cultures of these microbes (Singh 1981). *Micrococcus* spp. were the most active in converting dieldrin to more polar metabolites. A mixed anaerobic population in enrichment cultures yielded pure cultures of *Clostridium bifermentans*, *Clostridium glycolium* and a *Clostridium* sp. that could dehalogenate dieldrin producing *syn-* and *anti*-monodechlorodieldrin (Maule et al. 1987). It could also dehalogenate lindane, endrin, aldrin, and dieldrin and several other cyclodienes. The metabolism of lindane was more extensive than that of the cyclodienes.

6. Halogenated Aromatics.   Halogenated aromatics are commonly found as pollutants in ground water. Their degradation by microbes has attracted some attention and a number of organisms have been isolated that degrade representative compounds. Vandenbergh et al. (1981) isolated a bacterium, *Pseudomonas cepacia*, from soil of a landfill area with a previous history as a disposal site for chlorinated organic wastes, and found that this organism could utilize 2,6-dichlorotoluene as a carbon source for growth and this plasmid-borne characteristic could then be transferred to a mutant of *Pseudomonas aeruginosa*.

A study of the effects of spatial and temporal environmental variations on the rates of biodegradation of various pollutants including chlorobenzene and 1,2,4-trichlorobenzene showed that freshwater had the highest heterotrophic uptake rate when compared with samples from estuarine and marine sites (Bartholomew and Pfaender 1983). In surface and groundwater samples incubated under anoxic conditions favorable for denitrification, no significant utilization of chlorinated benzenes, ethylbenzene, and naphthalene was detected (Bouwer and McCarty 1983).

Reineke and Knackmuss (1984) used chemostat enrichment with an inoculum of mixed soil and sewage samples to obtain a bacterium that could use chlorobenzene as a carbon source. The initial enrichment substrate was benzene and this was gradually replaced by chlorobenzene. The bacterium degraded chlorobenzene via 3-chlorocatechol, 2-chloro-*cis,cis*-muconate, and 3-oxoadipate. The bacterium, designated NR1306, was not characterized.

The metabolism of 4-chloronitrobenzene by the yeast *Rhodosporidium* sp. led to the production of 4-chloroacetanilide and 4-chloro-α-hydroxyacetanilide following the intermediate production of 4-chloronitrosobenzene, 4-chlorophenylhydroxylamine, and 4-chloroaniline (Corbett and Corbett 1981).

The isolate of DeBont et al. (1986), obtained from mixed soil and water samples, able to use 1,3-dichlorobenzene as a sole carbon and energy source, was tentatively designated an *Alcaligenes* sp. Metabolism of the 1,3-dichlorobenzene was considered to proceed via 3,5-dichlorocatechol and 2,4-dichloromuconate with subsequent dehalogenation steps. Activated sludge was the source of the *Pseudomonas* sp. isolated by Spain and Nishino (1987) for its ability to utilize 1,4-dichlorobenzene as a sole source of carbon and energy. Major intermediates in the degradation of the dichlorobenzene were identified as 3,6-dichlorocatechol and 2,5-dichloro-*cis,cis*-muconate. The bacterium dehalogenated *o*-, *m*-, and *p*-dichlorobenzene as well as 2,5-dichlorophenol, 4-chlorophenol, 3-chlorocatechol, 4-chlorocatechol, and 3,6-dichlorocatechol.

7. Hydrocarbons.   As some of the polynuclear aromatic hydrocarbons represent chemical structures found also in various pesticides, information on their biodegradation is relevant. The findings of Ehrlich et al. (1982) on the degradation of naphthalene and Bauer and Capone (1985) on the degradation of anthracene and naphthalene showed that biodegradation of these polynuclear aromatic hydrocarbons did not occur under anaerobic conditions. The removal of trace organic compounds such as α-methylnaphthalene in wastewater by rapid infiltration of soil columns was inhibited if flooded and the soil columns became anaerobic (Hutchins et al. 1984b).

Zeyer et al. (1986b) found that *m*-xylene could be oxidized to carbon dioxide by microbes in a laboratory aquifer column operated under conditions favoring denitrification. Toluene was also oxidized by denitrifying organisms that were adapted to *m*-xylene. Aromatic hydrocarbons need not necessarily be persistent in anoxic environments such as sediments, sludge, and groundwater infiltration zones. Transformation of 1- and 2-methylnaphthalene by *Cunninghamella elegans* to 1- and 2-hydroxymethylnaphthalene was described by Cerniglia et al. (1984).

B. Organophosphates

Organophosphate insecticides have been in use for more than 30 yr and the role of microbes in their degradation has been well established. Lal (1982) and Barik (1984) have reviewed the literature on their metabolism. Lal (1982) cautions that it would be unwise to view the metabolism of organophosphates by microbes in isolation with disregard for the modifying effects that chemical, photochemical, and physical factors might have on the final outcome of biodegradation. The interplay of these factors may lead to quite different metabolic routes. The fate of malathion and parathion in the environment has been reviewed by Mulla et al. (1981).

1. Chlorpyrifos.   The phosphorothioate insecticide chlorpyrifos has a broad range of activity. Leoni et al. (1981) detected the formation of 3,5,6-trichloro-2-pyridinol in treated soil. Very little residual chlorpyrifos was detected after 6 mon. The half-life of chlorpyrifos in a silt-loam was 12 wk compared with 4 wk in a clay-loam (Getzin 1981). The breakdown of ^{14}C-chlorpyrifos in the clay loam in a 32 wk period led to 47% mineralization of the ^{14}C. The half-life of chlor-pyrifos in the estuarine environment was found to be 24 d (Schimmel et al. 1983). Chlorpyrifos was found to be more stable in sterilized samples of a sandy loam and a muck soil than in nonsterilized samples of the same soils (Miles et al. 1984).

2. Coumaphos.   Kearney et al. (1986) have studied the effectiveness of com-bined microbial activity and UV-ozonation in the degradation of coumaphos in the wastewater from livestock dips. The microbe used was a *Flavobacterium* sp., able to cleave the phosphorothioate linkage to give chlorferon (3-chloro-4-methyl-7-hydroxy coumarin) and diethyl thiophosphoric acid. No oxidation of the benzene ring was detected. However, microbial metabolism followed by UV-ozonation was effective in cleaving both the phosphorothioate linkage and destroying the chloroferon produced by microbial activity. A major ring-containing product gave GC-MS data characteristic of 2,4-dihydroxyacetophe-none. Addition of the coumaphos dip solution after it had been subjected first to microbial metabolism then UV-ozonation to a silty clay loam resulted in the release of aromatic ring ^{14}CO$_2$ indicating extensive degradation of the products from coumaphos pretreatment by the soil microflora.

3. Diazinon.   Diazinon was rapidly degraded in soil that had previously received 14 field applications of the insecticide over a 3-yr period (Forrest et al. 1981). This activity could be transferred to other soils by "inoculation" of these soils with some of the adapted soil. A *Flavobacterium* sp. that could degrade diazinon was isolated. Investigations by Merritt et al. (1981) of a green discolora-tion of fleece-rot and fly-strike lesions of sheep led to the isolation of a culture of

Pseudomonas aeruginosa that was able to degrade diazinon, used to protect sheep from blowfly. The bacterium could also degrade fenthion ethyl. Two isolates from flooded soil, a *Flavobacterium* sp. and a *Pseudomonas* sp., were found to hydrolyze diazinon (Adhya 1981b).

4. Fenitrothion and Fenitrooxon.   Adhya et al. (1981a) have recorded the formation of aminofenitrothion from fenitrothion in flooded acid sulfate soils, whereas little degradation was recorded for fenitrothion in samples of the same soils in which aerobic processes were dominant. The presence of hydrogen sulfide in the flooded soils aided the formation of desmethylaminofenitrothion through dealkylation reactions. Fenitrothion was found to be degraded via hydrolysis in soils of more positive redox potential, whereas in soils with a more negative redox potential, the degradation process included nitro-group reduction. Adhya et al. (1981b) isolated a *Flavobacterium* sp. from flooded soil that hydrolyzed fenitrothion.

Studies employing aquatic microcosms (Weinberger et al. 1982b) showed that fenitrothion was degraded more extensively in estuarine water microcosms than in lake water and distilled water microcosms. Desmethyl fenitrothion and 3-methyl-4-nitrophenol were major degradation products. In an additional study (Weinberger et al. 1982a), using both field and laboratory microcosms incubated in the light, major degradation products were fenitrooxon, demethylfenitrothion, *S*-methylfenitrothion, carboxyfenitrothion, and carboxyaminofenitrothion. Only hydrolysis products were detected in microcosms incubated in the dark. The degradation of fenitrothion appeared to be photocatalyzed. Evidence for biological degradation of fenitrothion and fenitrooxon was provided by Baarschers and Heitland (1986), who found that the fungus *Trichoderma viride* could hydrolyze both compounds to 3-methyl-4-nitrophenol which was then further degraded by cometabolic reactions. They expressed the opinion that the ability of the fungus to cometabolize the phenolic intermediate was of greater environmental relevance than bacteria described for their ability to use such organophosphorus compounds as sole sources of carbon, phosphorus, and nitrogen.

5. Fensulfothion.   Degradation of fensulfothion by a mixed culture of soil organisms was reported by Miles and Moy (1982). The conversion of fensulfothion to fensulfothion sulfide by *Klebsiella pneumoniae* was sensitive to molecular oxygen (Timms and MacRae 1982). The sulfide was rapidly bound by both living and dead cells of *K. pneumoniae* and concentrated in the cell membrane. The ability to reduce fensulfothion to its sulfide was found to be a feature of a diverse array of microbes (Timms and MacRae 1983) including a *Hafnia* sp. isolated from soil (MacRae and Cameron 1985). Sheela and Pai (1983) isolated two bacteria, *Pseudomonas alcaligenes* and an *Alcaligenes* sp. from soil that could use fensulfothion as a carbon source. The mode of attack on the fensulfothion molecule by *P. alcaligenes* appeared to be via hydrolysis to yield *p*-methylsulfinylphenol and diethylphosphorothioic acid. The *p*-methylsulfinylphenol was not utilized by the bacterium. Oxygen-limited cultures of *K. pneumoniae* reduced 4-methylsulfinyl phenol, an hydrolysis product from fensulfothion to 4-methylthiophenol (MacRae and Cameron 1985).

6. Methyl parathion.   Degradation of methyl parathion in flooded acid sulfate soils proceeded via aminomethylparathion and if the hydrogen sulfide formed by sulfate reduction was abundant, dealkylation yielded desmethylaminomethyl parathion (Adhya et al. 1981a). Low redox potential in a flooded soil favored degradation by nitro-group reduction whereas when applied to a soil that had a more positive potential, methyl parathion underwent hydrolysis (Adhya et al. 1981c).

The mineralization of methyl parathion was rapid at soil-water contents equivalent to 110 and 33 kPa (Ou et al. 1983). Under these conditions the pesticide appeared to undergo hydrolysis to *p*-nitrophenol which was then reduced to *p*-aminophenol. Mineralization of methyl parathion was greatly reduced at a soil-water content near saturation (Ou 1985). At a soil-water content more than 6 kPa, *p*-nitrophenol and *p*-aminophenol were the major metabolites.

As a contribution to the very sparse information on the persistence of pesticides in estuarine environments (Schimmel et al. (1983) found that methyl parathion had a half-life of only 1.2 d. Faster degradation took place in estuarine sediment and water column ecocores than in shake flasks inoculated with sediment plus water or water alone (Van Veld and Spain 1983). The ecocore was able to provide the aerobic and anaerobic gradients and possibly more nutrients required by the methyl parathion-degrading population. Such a population may well have been a consortium of species.

Aufwuchs bacteria, that is bacteria that are in attached growth, mats, or streamers in aquatic environments, were found to bring about a rapid transformation of methyl parathion (Lewis and Holm 1981). Of the several bacterial cultures isolated from aufwuchs that rapidly transformed methyl parathion, *Flavobacterium aquatile* and *Staphylococcus saprophyticus* were identified.

Lewis et al. (1984) examined the interaction of culture filtrates, mixed populations and some microbial products on the transformation of methyl parathion by bacteria. They found inhibition, which was related to treatments that lowered the pH, and stimulation which was related to an increase in viable counts. These may be some of the complicating factors also affecting the degradation of methyl parathion by *Flavobacterium* sp. in pure culture where the transformation of the insecticide was found to follow multiphasic kinetics (Lewis et al. 1985).

7. Parathion.   Parathion is highly toxic and has been used as an insecticide and an acaricide. Fuhremann and Lichtenstein (1980) detected the formation of paraoxon, aminoparathion, *p*-nitrophenol, and *p*-aminophenol in a soil-oat plant system that had been treated with parathion. Presumably these substances were produced at least in part by microbial activity. During an investigation to discover why diazinon failed to control an attack of root aphids on lettuce in a soil that had received 14 applications of diazinon over a 3-yr period, Forrest et al. (1981) were able to isolate a *Flavobacterium* sp. from the soil that could hydrolyze diazinon and parathion. When added to cell suspensions of the bacterium, the hydrolysis of parathion to *p*-nitrophenol was more rapid than the hydrolysis of diazinon and yet in the field, diazinon was rapidly degraded but parathion was not. The authors suggested that differences in the reaction mechanisms might be the reason for these unexpected results.

Adhya et al. (1981b) found that a *Flavobacterium* sp. and a *Pseudomonas* sp. could hydrolyze both parathion and diazinon. The presence of glucose inhibited the hydrolysis of parathion by *Pseudomonas* sp. but not by *Flavobacterium* sp., evidence that the hydrolases involved differ. The hydrolysis product from parathion, 4-nitrophenol, was not metabolized by *Flavobacterium* sp. but was converted to the 4-aminophenol by the pseudomonad. Adhya et al. (1981c) found no effect of varying the conditions that affect redox potential in a flooded soil on the degradation of parathion. Under all conditions imposed on the soil, degradation proceeded via nitro-group reduction. Adhya et al. (1981a) also reported the degradation of parathion to aminoparathion, in a group of five flooded soils which included two acid sulfate soils. The aminoparathion underwent dealkylation to desethyl aminoparathion only in soils with high sulfate content. Presumably, sulfate reduction in the soil provided the hydrogen sulfide involved in the dealkylation.

Mineralization of parathion in soils amended with various fertilizers, captafol, and atrazine was found to be inhibited by all amendments (Lichtenstein et al. 1982). The authors suggested that the reduced mineralization was due to effects on soil microbes that are responsible for the degradation of parathion, notably the soil fungi. The presence of rice plants in flooded soil was found to stimulate the mineralization of parathion ring carbon (Reddy and Sethunathan 1983). Mixing parathion with the soil rather than surface application led to a very significant increase in its persistence (Lichtenstein et al. 1983). It also led to decreased mineralization of parathion. The most notable effect was obtained when soil was also flooded. Mineralization practically ceased and up to 73% of the pesticide was bound. While nitro-group reduction was the major means of degradation in a group of saline soils, parathion was more persistent in saline soils and this was related to low soil microbial activity (Reddy and Sethunathan 1985).

A laboratory pilot plant study on parathion degradation in an activated sludge revealed its conversion to aminoparathion (Laplanche et al. 1981). Nelson et al. (1982) found that hydrolysis of parathion in a soil was due largely to microbial activity, and Nelson (1982) was able to isolate a *Bacillus* sp. from the soil with weak hydrolytic activity in culture. An *Arthrobacter* sp. isolate had strong activity in culture and could also effect the rapid hydrolysis of the pesticide in sterilized parathion-treated soil. *Arthrobacter* sp. could utilize parathion or *p*-nitrophenol as sole carbon source for growth.

Parathion hydrolase activity by *Pseudomonas diminuta* (Serdar et al. 1982) and *Flavobacterium* sp. (Mulbry et al. 1986) was found to be plasmid-borne.

8. Glyphosate.   Glyphosate is a potent, broad-spectrum herbicide that has been popular especially as it is considered to be rapidly inactivated in soil. However, 120 d after a soil application of the herbicide, Eberbach and Douglas (1983) were able to detect reduced nitrogenase activity, nodule numbers, and root weight in an indicator plant, subterranean clover. Although no chemical analyses were performed, the authors assumed that the toxicity to the indicator plant resulted from residual glyphosate in the soil.

Moore et al. (1983) isolated a *Pseudomonas* sp., strain PG 2982, that utilized glyphosate as a sole source of phosphorus. A later study by Shinabarger et al. (1984) showed that strain PG 2982 could also use a range of other organophosphonate compounds as phosphorus source. Jacob et al. (1985) described the complete pathway for glyphosate degradation by strain PG 2982 and found that the organism cleaved the phosphonomethyl carbon-nitrogen bond of glyphosate, producing glycine. Balthazor and Hallas (1986) isolated a *Flavobacterium* sp. from an activated sludge plant that could utilize glyphosate as a sole source of phosphorus. This bacterium could degrade glyphosate in the presence of orthophosphate which is of practical significance if the organism is to be of importance in natural environments or possibly in clean-up procedures. Recent studies with a glyphosate-degrading *Arthrobacter* sp. (Pipke et al. 1987) revealed that orthophosphate and organophosphates suppressed the glyphosate uptake system.

9. Miscellaneous Organophosphates.   Although Forrest et al. (1981) were unable to detect biodegradation of chlorfenvinphos in four loam soils, evidence for soil biodegradation of this pesticide in a loam and muck soil was obtained by Miles et al. (1984). The *cis*-isomer of chlorfenvinphos disappeared more rapidly than the *trans*-isomer.

The insecticide dichlorvos is used widely in crop protection and as a household and public health fumigant. In a study of dichlorvos decomposition in an enrichment culture of sewage microbes, Liebermann and Alexander (1983) proposed a pathway for its degradation which included the formation of dichloroethanol, dichloroacetic acid, and ethyl dichloroacetate.

A *Pseudomonas aeruginosa* strain isolated from fleece-rot lesions of sheep was reported to have the ability to degrade fenthion ethyl but no information on products or the extent of degradation was obtained (Merritt et al. 1981).

Malathion was used by Paris et al. (1981) in a study of the kinetics of microbial degradation of pesticides in water. They found degradation rates to be proportional to bacterial numbers and pesticide concentration.

The microbial degradation of the insecticide and acaricide phorate in soil led to the formation of phorate sulfone and phorate sulfoxide in a loam soil and a sandy soil, but the formation was much less in the sandy soil (Fuhremann and Lichtenstein 1980). Chapman et al. (1982b) also detected the formation of the sulfoxide and sulfone in a sandy loam and a muck soil and concluded that biochemical processes played a major part in the transformations of phorate, the sulfoxide and sulfone in soil. However, Forrest et al. (1981) did not detect degradation of phorate in sandy loam soils.

Chapman et al. (1982a) concluded after comparing the persistence of the phosphorodithioates terbufos, terbufos sulfoxide, and terbufos sulfone in sterile and nonsterile soil that biochemical processes were of major importance in the transformation of terbufos and its oxidation products in soil. Chapman et al. (1986c) showed that the insecticide isofenphos also disappeared faster in previously treated soil than in untreated soil.

C. Carbamates

Representative carbamates include insecticides, nematicides, herbicides, and fungicides and are used extensively, often as replacement pesticides for some of the persistent halogenated hydrocarbons and derivatives. The persistence of various carbamate pesticides in soil has been reviewed by Rajagopal et al. (1984a).

1. Carbaryl. The fate of carbaryl in the environment has been reviewed by Mount and Oehme (1981) and Barik (1984). Pretreatment of flooded soils with carbaryl led to more rapid degradation of subsequent additions of the pesticide (Rajagopal et al. 1983). Since steam sterilization of enrichment cultures from the soils led to the loss of carbaryl-degrading activity, biodegradation of the insecticides in the soils is indicated. Enrichment cultures obtained from carbaryl and carbofuran amended soils were able to degrade carbofuran and carbaryl respectively, indicating that the primary hydrolysis step is common to both pathways. Additives of nitrogen in the form of ammonium sulfate or urea increased the persistence of carbaryl in flooded soils with a low nitrogen content but not in an alluvial soil with a higher nitrogen content (Rajagopal and Sethunathan 1984). Also, hydrolysis of carbaryl seemed to be the major pathway of degradation because 1-naphthol accumulated. Hydrolysis was found to be the major course for microbial degradation of carbaryl in soil enrichment cultures and cultures of a *Bacillus* sp. with 1-naphthol and 1,4-naphthoquinone accumulating in the medium (Rajagopal et al. 1984b). Pretreatment of a flooded soil with 1-naphthol led to more rapid disappearance of carbaryl added subsequently to the same soil, but the accumulation of 1-naphthol and the formation of bound residues was greater in the pretreated soil (Rajagopal et al. 1986).

The biodegradation of carbaryl in soil can be affected by pH. Larkin and Day (1985) found that when continuous enrichment cultures were maintained at pH

6.0 for more than 2,000 hr they were unable to select a single bacterium that could degrade carbaryl. However, their labor was rewarded with a bacterial community of at least 12 to 14 members that could use carbaryl as a sole carbon and nitrogen source. When the pH of the soil perfusion columns and continuous enrichment culture was lowered to pH 5.2 and 5.0, respectively, they were able to isolate a *Pseudomonas* sp. that could use carbaryl as a sole carbon and nitrogen source. Larkin and Day (1986), with soil perfusion column enrichments, further illustrated the effect of pH on carbaryl degradation, yielding two isolates, *Pseudomonas* sp. (NCIB12042) and *Rhodococcus* sp., that used carbaryl as a sole carbon and nitrogen source at pH 6.8 The carbaryl was metabolized slowly. An enrichment maintained at pH 5.2 yielded another isolate, *Pseudomonas* sp. (NCIB 12043), that degraded carbaryl rapidly. The strain NCIB 12042 metabolized 1-naphthol via salicylic acid and the *Rhodococcus* sp. produced salicyclic acid and gentisic acid.

Degradation of carbaryl in low nutrient aquatic environments was concluded to occur by abiotic processes, whereas in polluted river water both abiotic and biotic processes were important (Liu et al. 1981b).

Carbaryl has been noted to stimulate methanogenesis in anaerobic salt marsh soils and organic-rich aquifer soils (Kiene and Capone 1986). The authors suggested that the monomethylamine formed by the microbial hydrolysis of carbaryl under anaerobic conditions serves as a substrate for methanogenic bacteria. Thomas et al. (1986) have reported that the mineralization of carbaryl by a mixed culture of microbes proceeded at a logarithmic rate.

2. Carbofuran.   Metabolites from carbofuran recovered from a loam and a sandy soil were 3-ketocarbofuran, 3-hydroxycarbofuran carbofuranphenol, 3-ketocarbofuranphenol, and 3-hydroxycarbofuranphenol (Fuhremann and Lichtenstein 1980). Repeated field applications of a carbofuran preparation over a 3-yr period led to enhanced degradation of carbofuran in soil and a *Pseudomonas* sp. was isolated that could degrade carbofuran (Felsot et al. 1981). Only traces of the insecticide were detected, 30 d after application. Soil with a history of phorate application, however, gave longer persistence of carbofuran with 8 to 21% of the insecticide detected after 30 d. Rapid disappearance of carbofuran and 3-ketocarbofuran from a loam soil was also found by Miles et al. (1981). Rajagopal and Sethunathan (1984) found that additions of ammonium sulfate to a flooded laterite soil lengthened the persistence of carbofuran. Hydrolysis was indicated as a major route of degradation of carbofuran because of the accumulation of 2,3-dihydro-2,2-dimethylbenzufuran-7-ol in the soil. Negligible mineralization of ^{14}C-ring labeled carbofuran occurred.

Soils amended with cow manure brought about increased mineralization of ^{14}C-labeled carbofuran and increased amounts of bound carbofuran residues (Koeppe and Lichtenstein 1984). When soils were amended with sterile, γ-irradiated manure, reduced bound residues and carbon dioxide were produced, indicating a role for manure organisms in the biodegradation.

A single pretreatment of a sandy loam soil with carbofuran greatly increased the rates of degradation of subsequent carbofuran additions (Harris et al. 1984). This enhanced activity was reduced by treatments aimed at destroying or reducing biological activity and was related to the amount of applied carbofuran and the frequency of application. The enhanced activity was not confined to carbofuran since accelerated rates were also found for other carbamates such as aldicarb, bufencarb, carbaryl, and cloethocarb. Further work by Chapman et al. (1986a) detected enhanced carbofuran degradation in a sand, sandy loam, clay-loam, and a muck soil within 28 d from the initial application. Carbaryl also disappeared rapidly from the same soils following carbofuran pretreatment. An extreme example of enhanced degradation was the carbofuran-pretreated sandy loam soil in which a further application of carbofuran was degraded to < 5% of the amount applied within 24 hr (Chapman et al. 1986b). By contrast, more than 95% of the applied carbofuran remained in a sample of the same soil that had not been pretreated.

Accelerated degradation of carbofuran in a flooded soil has also been described (Rajagopal et al. 1986). In this case, the pretreatment of the soil was with the hydrolysis product, 2,3-dihydro-2,2-dimethylbenzofuran-7-ol and not the pesticide. Studies on the metabolism of carbofuran by pure culture of soil isolates have shown that 2,3-dihydro-2,2-dimethylbenzofuran-7-ol is a major metabolite in the case of *Bacillus* sp. (Rajagopal et al. 1984b), *Azospirillum lipoferum* and *Streptomyces* spp. (Venkateswarlu and Sethunathan 1984), and *Pseudomonas cepacia* and *Nocardia* sp. (Venkateswarlu and Sethunathan 1985). The 3-ketocarbofuran and 3-hydroxycarbofuran were also produced by the *Bacillus* sp. (Venkateswarlu and Sethunathan 1984) and the 3-hydroxycarbofuran was detected in cultures of *Azospirillum lipoferum* and *Streptomyces* spp. (Venkateswarlu and Sethunathan 1984). An unidentified colored compound was produced by *Nocardia* sp. during the degradation of carbofuran (Venkateswarlu and Sethunathan 1985).

3. Miscellaneous Carbamates.   In a study of the degradation of the carbamates desmedipham, phenmedipham, promecarb, and propamocarb, a group that includes two herbicides, an insecticide and a fungicide, Knowles and Benezet (1981) found that the ones with two aromatic rings in their structure were most susceptible to degradation by pure cultures of eight bacteria, two fungi, and a yeast. These were the herbicides desmedipham and phenmedipham. Propamocarb, an aliphatic fungicide, did not appear to be susceptible to attack by any of the cultures, while promecarb with a single aromatic ring was intermediate in biodegradability.

Cycloate, a carbamate herbicide, was not mineralized to any significant extent in sewage or lakewater, but it was cometabolized to organic products (Novick and Alexander 1985).

The fate of aldicarb [α-methyl-α-(methylthio)propionaldehyde-*O*-(methylcarbamoyl)oxime] in soil and water has been studied. Ou et al. (1985) investigated its degradation in eight soils and found that in general, the *S*-methyl-^{14}C aldicarb was mineralized faster in surface soils incubated under aerobic conditions than in subsurface soils incubated either aerobically or anaerobically. They detected aldicarb sulfoxide, aldicarb sulfone, aldicarb sulfoxide oxime, aldicarb sulfoxide nitrile, and aldicarb sulfone, aldicarb sulfoxide oxime, aldicarb sulfoxide nitrile, and aldicarb sulfone oxime as metabolites. The fate of aldicarb, the sulfoxide, and sulfone in aerobic and anaerobic groundwater was examined by Miles and Delfino (1985). Little oxidation of aldicarb occurred but reduction of the sulfoxide to aldicarb was significant. They concluded that the fate of the three compounds in Floridian groundwater is complicated.

Kiene and Capone (1986) reported the stimulation of methanogenesis in anaerobic salt marsh soil and an aquifer soil that had been treated with the insecticide methomyl. Monomethylamine was probably produced by biological hydrolysis of methomyl, and methane production was stimulated as a result of utilization of the amine by methanogenic bacteria.

The blue-green bacterium, *Anacystis nidulans*, hydrolyzed the preemergence herbicide CIPC to 3-chloraniline, possibly by the activity of an acylamidase (Wright and Maule 1982). Vega et al. (1985) isolated a strain of *Pseudomonas cepacia* from soil that could utilize CIPC and its degradation product, 3-chloroaniline, as sole sources of carbon and energy. Yeast extract addition to the medium increased the rate of CIPC degradation. The bacterium retained its ability to degrade CIPC even after seven subcultures into a medium without the herbicide. Studies of the mineralization of chlorpropham in natural waters showed that the rate was proportional to bacterial numbers and pesticide concentration (Paris et al. 1981).

Wang et al. (1984) found that the herbicide propham was mineralized at concentrations of 400 pg and 1 µg/mL in sewage samples but only at 400 pg/mL in lakewater. At 1 µg/mL, the propham was converted to organic products, presumably by cometabolism. The authors stressed the need for testing to determine whether a substance giving a negative mineralization test is converted to organic products. Further studies on the rates of mineralization of propham by Hoover et al. (1986) revealed that some pollutants such as propham were not mineralized in lakewaters at 1 µg/mL. However, at 10 ng/mL or less, propham was mineralized although not in all lakewater samples. Hoover et al. (1986) suggested that the use of a high concentration of a chemical in laboratory biodegradation tests may not provide data meaningful for natural environments with much lower concentrations of the chemical.

D. Thiocarbamates

1. Diallate and Triallate.   The biodegradation of two herbicides, diallate and triallate, in soil proved to be directly related to the microbial biomass (Anderson 1984). Addition of glucose or a mixture of finely divided cellulose and plant residues to the soil enhanced the release of $^{14}CO_2$ from both herbicides. The major product from carbonyl-^{14}C-labeled diallate degradation was ^{14}CO$_2$ whereas allyl-2-^{14}C-labeled triallate was degraded chiefly to a ^{14}C-soil residue.

2. EPTC.   Lee (1984) reported the disappearance of the preemergence herbicide EPTC from EPTC-amended nutrient broth cultures of nine bacterial isolates including *Micrococcus* spp., *Pseudomonas* spp., *Alcaligenes* spp., and *Bacillus* spp. Twenty-nine fungal isolates were also reported to degrade the herbicide during growth in a complex, rich medium. No information on degradation products was obtained. Since the bacterial cultures lost the ability to degrade EPTC during a 15-mon storage period, Lee (1984) suggested that plasmids may carry the genes for EPTC degradation. Tam et al. (1987) were able to demonstrate that this is the case for an *Arthrobacter* sp. strain TE1 that was isolated from an EPTC-treated soil. The *Arthrobacter* sp. could grow on EPTC as a sole carbon source. They found four plasmids harboured by the bacterium and a high frequency of spontaneous mutation associated with loss of EPTC degradation ability.

E. Triazines

The degradation of triazine herbicides, a relatively persistent group, has been discussed by Cook and Hütter (1981b).

The rapid and complete bacterial degradation of the s-triazines, cyanuric acid, ammelide, ammeline, melamine, and four *N*-alkylated ammelides and ammelines by two strains of *Klebsiella pneumoniae* and three *Pseudomonas* spp. isolated from sewage and soil has been reported (Cook and Hütter, 1981a). Specific growth rates of the bacteria with the s-triazines as sole nitrogen source were similar to those obtained with ammonium nitrogen.

The compound that represents the basic chemical structure for the s-triazine herbicides, 2-chloro-1,3,5-triazine-4,6-diamine, has been shown by Grossenbacher et al. (1984) to serve as a sole nitrogen source for a *Pseudomonas* sp. The bacterium utilized the amine group in the 6-position with the production of the acid-labile, 2-chloro-4-amino-1,3,5-triazine-6(5H)-one.

Deethylsimazine, considered to be an early product in the microbial metabolism of chloro-s-triazine herbicides, was converted to 1 mol each of 6-(ethylamino)-1,3,5-triazine-2,4(1H,3H)-dione, chloride ion, and ammonium ion by a soil isolate identified as *Rhodococcus corallinus* (Cook and Hütter 1984). The bacterium utilized only one of the nitrogens of the deethylsimazine molecule as a source of nitrogen for growth. When this bacterium was grown together with a strain of a *Pseudomonas* sp., all five nitrogens of deethylsimazine were assimilated by the bacteria.

1. Atrazine.   The degradation of the pre- and postemergence herbicide atrazine by a *Nocardia* sp. has been described by Giardina et al. (1982). *N*-Dealkylation of the atrazine molecule yielded 4-amino-2-chloro-1,3,5-triazine as the major route of metabolism by the actinomycete. Jessee et al. (1983) were able to isolate a facultatively anaerobic bacterium that could degrade atrazine anaerobically.

N-Dealkylation of atrazine proved to be an important reaction in its degradation by three *Pseudomonas* spp. that could use the pesticide as a sole source of carbon (Behki and Khan 1986). The process yielded deisopropylatrazine. Since Cook and Hütter (1984) showed that this substance, termed deethylsimazine, could be completely degraded by a mixed culture of *Rhodococcus corallinus* and *Pseudomonas* sp., Behki and Khan reasoned that complete degradation of atrazine in soil was possible.

2. Metamitron. Degradation of the triazinone herbicide metamitron by an *Arthrobacter* sp. has been demonstrated with the formation of benzoylformic acid acetylhydrazone and benzoyl formic acid as major metabolites (Engelhardt et al. 1982). The degradation proceeds via an hydrolytic cleavage of the triazinone ring between N4 and C5.

3. Prometryn and Ametryn. Cook and Hütter (1982) were successful in the isolation of nine pure cultures of bacteria that could utilize prometryn and ametryn or both as sole sources of sulfur for growth. The bacteria were not fully characterized but one of the strains had the major characteristics of a fluorescent pseudomonad. Utilization of the sulfur of the two herbicides resulted in the formation of the corresponding hydroxy derivatives. It would be interesting to test whether the organisms could also utilize the sulfur of other methylthio-substituted compounds such as methylthiophenol, a product of the hydrolysis of a number of organophosphates.

F. Phenoxy Compounds

The fate of phenoxy herbicides in the forest ecosystem has been reviewed by Norris (1981). This group is widely used and much has been published on the biodegradation of its representatives. Nevertheless, there are still problems of persistence of some in soil and water. For example, very little work has been done on their fate in anaerobic environments. Liu et al. (1981a) described a cyclone fermenter test system that can be used to determine the biodegradability of pesticides and other substances under a variety of environmental conditions. They have used this apparatus to test the biodegradability of 2,4-D under aerobic and anaerobic conditions, with and without cometabolites. With a mixed inoculum derived from activated sludge, soil and sediments, they found that while 2,4-D was easily degradable under aerobic conditions, the degradation under anaerobic conditions was very much slower. Shaler and Klecka (1986) investigated the effect of dissolved oxygen on the degradation of 2,4-D by an enrichment culture of 2,4-D-degrading bacteria and found that oxygen concentrations less than 1 mg/L may be rate limiting for 2,4-D biodegradation. Using a two-substrate enrichment technique, in which either MCPA or dichlorprop was supplied plus either vanillic acid or benzoic acid, Kilpi (1980) was able to isolate mixed bacterial cultures from a soil previously treated with mecoprop that could use 2,4-D and MCPA as sole carbon sources. They chose vanillic acid as one of the nonherbicide substrates because its structure resembles that of MCPA, and benzoic acid was chosen since it is a common metabolite in the degradation of aromatic compounds in soil. During incubation of the enrichments, the nonherbicide molecule was attacked first. When benzoic acid was supplied as the cosubstrate, one of the mixed cultures was also able to degrade dichlorprop.

While it was not possible to detect directly 2,4,5-T-degrading organisms from waste dumping sites, after long chemostat culture in which 2,4,5-T was supplied at low concentration (50 µg/mL) and nonpesticide plasmid substrates such as toluate, salicylate, or chlorobenzoate at high concentration (250 µg/mL) Kellogg et al. (1981) were successful in obtaining a microbial population that could degrade 2,4,5-T. This population yielded a pure culture of *Pseudomonas cepacia* that could utilize 2,4,5-T as a sole carbon source (Kilbane et al. 1982). The bacterium could also utilize a variety of chlorinated phenols. Resting cells of the bacterium could effectively dehalogenate *o-*, *m*, and *p-* halogen monosubstituted phenols but not those with two chlorine atoms in the one position such as 2,3,6- and 3,4,5-trichlorophenols (Karns et al. 1983b). The authors suggested that 2,4,5-trichlorophenol is an intermediate in the degradation of 2,4,5-T. The bacterium could also dehalogenate various brominated and fluorinated phenols, but dehalogenation activity with iodinated phenols was weak.

Further work with this isolate indicated that the first step in 2,4,5-T degradation, the formation of 2,4,5-trichlorophenol by cleavage of the ether linkage, is catalyzed by constitutive enzymes, whereas inducible enzymes are involved in the degradation of 2,4,5-trichlorophenol (Karns et al. 1983a). Enzymes for the conversion of 2,4,5-trichlorophenol were repressed by the presence of other carbon sources. Chatterjee et al. (1982) found that inoculation of soil amended with 2,4,5-T at 1 mg/g with a pure culture of *Pseudomonas cepacia* was successful in that the bacterium multiplied and degraded up to 95% of the herbicide within a week. When heavily contaminated soil having 2,4,5-T up to 20 mg/g was inoculated with the bacterium and incubated to allow 2,4,5-T degradation, it was able to support the growth of 2,4,5-T-sensitive plants (Kilbane et al. 1983). Once the 2,4,5-T was reduced by biodegradation to undetectable levels, the population of the *Pseudomonas cepacia* declined rapidly.

Studies on the decomposition of uniformly ring-labeled ^{14}C-2,4,5-T in six soils by McCall et al. (1981) gave an average decomposition time of 14 d. The two degradation products, 2,4,5-trichlorophenol and 2,4,5-trichloroanisole, were detected in the soils. Smith and Hayden (1981) detected rapid degradation of MCPA, MCPB, and mecoprop in three soil types giving half-lives of less than 7, 6, and 8 d, respectively. MCPA was detected as a degradation product from MCPB in all moist soils. Soulas et al. (1983) found that pretreatment of soil with either 2,4-D or MCPA did not enhance the ability of the soil to degrade 2,4,5-T. In one of the rare studies on the anaerobic degradation of this group, Suflita et al. (1984) investigated the degradation of 2,4,5-T by an anaerobic consortium capable of growing on 3-chlorobenzoate. Unpredictably, the activity of the consortium led to the removal of the *p*- chlorine atom of the 2,4,5-T molecule rather than the *m*- chlorine and yielded 2,5-dichlorophenoxyacetic acid. Reductive dechlorination of the 2,4,5-trichlorophenol, resulting from cleavage of the ether linkage of 2,4,5-T by anaerobic sludge microbes, led to the formation of 3,4-dichlorophenol and 4-chlorophenol which persisted in the sludge (Mikesell and Boyd 1985).

Lillis et al. (1983) isolated a strain of *Pseudomonas putida* from soil that produces a primary alkylsulfatase that cleaves the O–S bond of 2-(2,4-dichlorophenoxy)ethyl sulfate giving rise to 2-(2,4-dichlorophenoxy)ethanol, which is the initial step in the activation of the nonphytotoxic parent material. Oxidation of the 2-(2,4-dichlorophenoxy)ethanol to 2,4-D by soil organisms completes the activation process. Enrichment cultures using inocula from river, stream, and pond water as well as activated sewage yielded 30 isolates capable of metabolizing 2,4-D (Amy et al. 1985). Members of the genus *Alcaligenes* were prominent among these isolates and one carried a plasmid of about 56 megadaltons (pEML159) that had characteristics very similar to those of a plasmid (pJMP397) isolated from a strain of *Alcaligenes eutrophus*. A cloned restriction fragment of 14.8 megadaltons expressed in *Escherichia coli* the ability to metabolize ^{14}C-acetate-labeled 2,4-D with the release of $^{14}CO_2$.

A microbial community comprised of two *Pseudomonas* spp., an *Alcaligenes* sp. a *Flavobacterium* sp., and *Acinetobacter calcoaceticus*, isolated from the rhizosphere of wheat plants, was shown to have the ability to grow on the herbicide mecoprop as a sole carbon source (Lappin et al. 1985). None of the bacteria could grow on the herbicide as pure cultures. The community could also degrade 2,4-D and MCPA.

The degradation of the phenoxy herbicides MCPA and 2,4-D in soil and the effects of various soil amendments on the degradation have been studied by Duah-Yentumi and Kuwatsuka (1982). Organic amendments such as rice straw and compost stimulated their degradation in soil perfusions. Of the three inorganic amendments, N, P, and K, only phosphorus enhanced the degradation of 2,4-D and MCPA.

Studying the effects of temperature and soil moisture on kinetics of 2,4-D degradation in a sandy loam, Parker and Doxtader (1983) found that the optimum temperature and moisture were 27°C and 0.1 bar, respectively. With increasing soil moisture tension, in the range 20 to 35°C, water availability to the 2,4-D-degrading population decreased and the concentration of 2,4-D in the soil solution increased, resulting in reduced 2,4-D degradation. The rates of mineralization in a chernozem soil of the two carbon atoms of the side chain of 2,4-D soil were found to be essentially similar except that the oxidation of the carboxyl carbon predominated during the early phases of incubation (Kunc and Rybarova 1983). Inoculation of a soil that had not previously been treated with 2,4-D with a soil suspension from a soil in which the numbers of 2,4-D degraders had increased led to accelerated mineralization of the side-chain carbon atoms.

The effects of sorption on the degradation of 2,4-D in soil have been studied by Ogram et al. (1985). Sorbed 2,4-D was completely protected from biological degradation, whereas sorbed and solution phase bacteria showed about the same efficiency for solution-phase 2,4-D degradation. Cells of *Alcaligenes eutrophus*, adsorbed to microscopic magnetic iron oxide particles, also were found to degrade 2,4-D in the solution phase in water (MacRae 1985).

Fournier (1980) devised a method for enumerating soil microbes that have the ability to metabolize or cometabolize 2,4-D. Cometabolizers were enumerated in a medium that contained ^{14}C-2,4-D plus a warm water extract of soil that provided the growth substrate(s) for the cometabolizing population. Only ^{14}C-2,4-D was provided as carbon source in the medium for 2,4-D metabolizers. Fournier et al. (1981) made use of this method to study the effect of 2,4-D application rates on the two kinds of 2,4-D-metabolizing populations in soil. They found a positive relationship between the amount of 2,4-D added to the soil and the size of the 2,4-D-degrading population, whereas the size of the cometabolizing population was unaffected by additions of 2,4-D to soil. Soulas and Fournier (1981) described a method used for the study of ^{14}C-2,4-D degradation in the soil. They chose 2 to 3 mm air-dried aggregates as the sampling unit and subjected them to various storage times and treatments. Prolonged air-dried storage of the soil aggregates lowered the ability of the 2,4-D-metabolizing population to develop. This was partially restored by rewetting. They emphasized that stored air-dried soil may not give a true reflection of microbial activity in a soil when it is used for experiments on microbial soil processes. Stott et al. (1983) followed the biodegradation, stabilization into humus, and incorporation into soil microbial biomass of ring-^{14}C and 2-^{14}C-side chain-labeled 2,4-D in four soils over a 1-yr incubation. Both ring- and side-chain-labeled 2,4-D were degraded rapidly and from 73 to 94% of the ^{14}C was released as ^{14}CO$_2$. After 1 year, from 3.3 to 15.3% of the ^{14}C activity was located in the soil microbial biomass. Soulas et al. (1984) also used a fumigation technique to study the degradation of ^{14}C-labeled 2,4-D in soil. The maximum amount of ^{14}C radioactivity incorporated into the soil biomass was 13.4%, irrespective of the location of the ^{14}C label.

Ou (1984) examined the effect of soil moisture tension and temperature on ring-labeled ^{14}C-2,4-D degradation in a loamy sand and a clay loam. The herbicide was rapidly mineralized in both soils when the tension was between 0.1 and 0.33 bar. The amount of 2,4-D carbon assimilated into the microbial biomass was estimated to be 4.2% of that applied.

Smith (1985a) compared the degradation of ring-labeled ^{14}C-2,4-D and 2-^{14}C side-chain-labeled 2,4-D in a clay loam, a clay and a sandy loam and found that both underwent rapid breakdown. Between 27 and 45% of the radioactivity was released as $^{14}CO_2$ within 10 d. Smith concluded that decarboxylation of 2,4-D did not occur. The initial step in 2,4-D degradation appeared to be cleavage of the ether linkage to 2,4-dichlorophenol which then underwent methylation to 2,4-dichloroanisole. Cleavage of the ether linkage in mecoprop followed by mineralization of the ring ^{14}C was found to occur in three soil types (Smith 1985b). The product of the ether cleavage, 4-chloro-2-methylphenol, was detected. Mecoprop degradation was rapid in all soils.

Pretreatment of soil with 2,4-D enhanced degradation of subsequent applications of 2,4-D and MCPA but not 2,4,5-T (Soulas et al. 1983). The authors considered that this confirmed that 2,4,5-T undergoes cometabolism. However, pretreatment of the soil with MCPA only enhanced 2,4-D degradation.

A study of the kinetics of microbial degradation of the butoxyethyl ester of 2,4-D in a large number of water samples showed that the rate of hydrolysis was proportional to both bacterial numbers and herbicide concentration (Paris et al. 1981).

Subba-Rao et al. (1982) found that 2,4-D at concentrations below 300 ng/mL was mineralized in lake water and sewage samples. The 2,4-D was mineralized only in samples from a eutrophic lake. They concluded that the metabolism of organic compounds such as 2,4-D at trace levels is different from that observed at higher concentrations. Preexposure of ecocores to 2,4-D at concentrations above 10 ng/mL led to adaptation of the microbial population and gave more rapid mineralization in preexposed ecocores. Below 10 ng/mL no adaptation was detected.

The transformation of the butoxyethyl ester of 2,4-D, by a model periphyton-dominated ecosystem (Lewis et al. 1983), showed the observed transformation rates were best predicted by using the ratio of colonized surface area to container volume. Measured transformation rates for butoxyethyl ester of 2,4-D by bacteria in the presence of other microorganisms and organic substances in some cases were inhibited and in some cases a stimulation was detected (Lewis et al. 1984). Amino acids, sugars, alcohols, and yeast extract additions to transformation mixtures led to increased bacterial transformation of the ester.

The degradation of 2,4-D was found to be six times faster under aerobic ($+500$ mV) than under anaerobic (-200 mV) conditions in a freshwater sediment (DeLaune and Salinas 1985). The rate of its degradation in the water column was almost six orders of magnitude less than the anaerobic sediment.

Gibson and Suflita (1986) have compared the degradation of 2,4-D and 2,4,5-T in samples from a variety of anoxic habitats including an anoxic aquifer, freshwater sediment, and sewage sludge. They found that 2,4-D was converted to 2,4-dichlorophenol, then dehalogenated in sludge, sediment, and a methanogenic aquifer. The pathway of 2,4,5-T degradation in sludge involved the formation of 2,4,5-trichlorophenol whereas in sediment, 2,4,5-T was dehalogenated at either the *p*- or *m*- position to give 2,5- or 2,4-dichlorophenoxyacetic acid.

Studies of 2,4-D mineralization rates in lakewater at concentrations from 100 pg/mL to 1 μg/mL led Hoover et al. (1986) to conclude that assessments of biodegradation of chemicals that are commonly carried out at relatively high concentrations may not be valid for the much lower concentrations usually found in natural environments.

Acclimation of activated sludge to 2,4-D degradation did not occur when the concentration of 2,4-D was as low as 1 μg/mL (Kim and Maier 1986). However, Wiggins et al. (1987) tested some of the hypotheses put forward to explain the acclimation period and found that when sewage samples were incubated for 16 d at 28°C before the addition of 2 ng/mL of 2,4-D, there was no acclimation period. They concluded that the acclimation period is mostly a reflection of the time it takes for the small population of active organisms to multiply and that this time may be lengthened by predation.

G. Phenols

1. Pentachlorophenol (PCP).   PCP is a multipurpose pesticide, used to control termites; to protect timber from fungal rots and wood-boring insects; and as an herbicide, defoliant, and disinfectant. Chemical contamination of lakes, streams, and groundwater is common and since it is an inhibitor of oxidative phosphorylation it is toxic to many organisms, and acutely toxic to many species of fish. Recently it was found to inhibit methanogenesis and could have a serious effect on the anaerobic digestion phase of waste water treatment (Guthrie et al. 1984).

Under most circumstances, PCP appears to be biodegradable. Pure cultures of *Arthrobacter* spp. (Stanlake and Finn 1982), *Flavobacterium* spp. (Saber and Crawford 1985; Steiert et al. 1987), *Pseudomonas* spp. (Trevors 1982), and uncharacterized bacterial strains (Pignatello et al. 1983) have been isolated that can degrade the pesticide. Dechlorination of PCP by *Pseudomonas cepacia* AC1100 has also been demonstrated (Karns et al. 1983b). The inducer of PCP metabolism by *P. cepacia* AC1100 appears to be 2,4,5-trichlorophenol (Karns et al. 1983a). Studies on the kinetics of growth on PCP of an enrichment culture of mixed bacterial and protozoan species revealed low maximum specific growth rate and yield coefficient (Klecka and Maier 1985). Growth was inhibited at a PCP concentration of 1.375 mg/L. Neilson et al. (1983) isolated two strains, both provisionally characterized as *Arthrobacter* spp., that carry out the methylation of PCP to pentachloroanisole.

Mineralization of PCP in a flooded soil has been demonstrated by Weiss et al. (1982), who concluded that PCP was not persistent under their prevailing soil conditions.

Attempts to remove PCP from PCP-contaminated soil by inoculation of soil with a PCP-degrading *Arthrobacter* sp. were successful (Edgehill and Finn 1983). Tests were carried out as laboratory incubation studies (30°C) and in an outdoor shed (8–16°C), and the rate of PCP disappearance was much more rapid at the higher controlled temperature. Trevors (1982) has reported the isolation of *Pseudomonas* spp. from soil that can degrade PCP at 4°C. Soil inoculation may prove useful in the decontamination of localized PCP contamination of soil.

PCP degradation in water and wastewater has also been studied in order to devise methods for its removal. In PCP-dosed artificial outdoor streams, photolysis in the surface water was the major means of PCP removal until the microflora of the streams had adapted (Pignatello et al. 1983). After a period of 3 wk, the microflora of the streams became the primary agent in PCP removal and mineralization occurred. The microbial populations in attached growth and surface sediment were the most important. Later studies with experimental outdoor streams showed that PCP was degraded both aerobically and anaerobically (Pignatello et al. 1986).

Mineralization of PCP in wastewater under aerobic conditions proceeded at a maximum rate when the concentration of PCP was around 350 µg/L (Moos et al. 1983). PCP in sewage sludge was degraded under anaerobic conditions (Guthrie et al. 1984). With acclimation of the digesters by gradually increasing the PCP in the influent, the digesters were found to cope with concentrations up to 5 mg/L without affecting methanogenesis. Acclimation of sewage sludge with monochlorophenols produced an anaerobic sludge that could degrade PCP (Mikesell and Boyd 1986). Degradation of PCP in anaerobic sludge samples was also demonstrated by Godsy et al. (1986) and Mikesell and Boyd (1985). Hannah et al. (1986) compared the efficiencies of six waste water treatment processes in the removal of toxic pollutants and found the activated sludge process gave best results.

Continuous culture studies using a *Flavobacterium* sp. and a natural consortium of epilithic microbes gave similar specific rates of PCP degradation (Brown et al. 1986). Because of the efficiency of PCP removal by the epilithic population, Brown et al. (1986) suggested that a fixed-film bioreactor could be used to remove PCP from contaminated water. Fixed-film bioreactors, consisting of a mixed microbial consortium as a biofilm on softwood bark, have been used successfully by Salkinoja-Salonen et al. (1983) and Apajalathi and Salkinoja-Salonen (1984) to remove PCP from water.

2. Phenol and Substituted Phenols.   Phenol is one of the most important pollutant chemicals because it is widely used as a structural precursor for many other classes of chemicals including agricultural chemicals, it is toxic to fish at low concentration, and it imparts objectionable flavor and odor to drinking water. Concentrations needed to produce these undesirable properties in water are extremely low, in the order of pg and ng/mL. This has presented rather special problems in degradation studies. Much research effort has therefore been directed to studies of its biodegradation in wastewater treatment and its removal from water.

The role of actinomycetes in the degradation of pesticides and other pollutants has received scant attention and Subba-Rao et al. (1982) have drawn attention to this deficiency. Antai and Crawford (1983) have studied the degradation of phenol by *Streptomyces setonii* in pure cultures held at 45°C. Catechol was formed and then the aromatic ring cleaved by catechol 1,2-dioxygenase to yield *cis,cis*-muconic acid.

Degradation of phenol by a continuous culture of *Pseudomonas putida* was enhanced when the organism was allowed to grow as a biofilm on stainless steel baffles in the culture vessel (Molin and Nilsson 1985). Phenol degradation rate was improved by a factor of about three over a culture without biofilm. While able to grow on *o*-nitrophenol as a sole carbon and nitrogen source, *Pseudomonas putida* B2 could not utilize a variety of para-substituted derivatives (Zeyer et al. 1986a). Enzymatic studies revealed that *o*-nitrophenol was the only compound tested that could act as inducer and substrate for the enzymes involved. In a study of phenol, phloroglucinol, *p*-cresol, and hydroquinone degradation in enrichment cultures developed using an inoculum from sewage sludge, Young and Rivera (1985) found the stoichiometric formation of CH_4 and CO_2. Dwyer et al. (1986) were successful in immobilizing a consortium of three physiological groups of bacteria and achieved stoichiometric conversion of the phenol to CH_4 and CO_2. The enrichment culture which consisted of a phenol-oxidizing bacterium, a *Methanothrix*-like bacterium and an H_2-utilizing bacterium, was immobilized in a long thin thread of 2% agar producing something similar to spaghetti. A major attribute of their immobilized culture was that the bacterial cells were protected from substrate inhibition by high phenol concentrations, resulting in the apparent K_i increasing from 900 to 1,725 µg/mL of phenol.

Studies on the anaerobic degradation of phenol and other chemical pollutants in sewage sludge are scarce and Ghisalba (1983) has stressed that there is still much to be done to improve the efficiency of waste treatment. Published works on phenol degradation in sewage sludge (Horowitz et al. 1982; Boyd et al. 1983) revealed the complete mineralization of phenol. The methanogenic degradation of phenol was found to be inhibited by the presence of pentachlorophenol at concentrations greater than 1 mg/L (Godsy et al. 1986).

Soils treated with 4-chloro-*o*-cresol and 5-chloro-3-methylcatechol at levels from 0.01 to 2 mg/g soil rapidly degraded both phenols (Sattar 1981), which are metabolites of the herbicide MCPA. Spain and Van Veld (1983) found that the mineralization of *p*-cresol was rapid in estuarine and marine ecocores. Preexposure of the microbial communities to *p*-cresol had little effect on the rate of mineralization. Phenol-acclimated activated sludge metabolized *o*-cresol via a number of pathways, reflecting the heterogeneity of the microbial population (Masunaga et al. 1986). GC-MS data confirmed the formation of three dihydroxytoluenes, 3-methylcatechol, 4-methyl resorcinol, and methylhydroquinone, during the degradation of the *o*-cresol.

Methanogenic degradation of phenols, including phenol and methylated phenols, has been suggested by Ehrlich et al. (1982) to occur in anoxic groundwater. Under anaerobic conditions prevailing in samples of an alluvial sand aquifer, the three cresol isomers were metabolized preferentially in the order *p-* > *m-* > *O-*(Smolenski and Suflita 1987). Sulfate-reducing conditions were more favorable for cresol degradation than methanogenic conditions. Degradation of the *p*-cresol appeared to proceed via hydroxylation of the methyl group. Suflita et al. (1987) later described the anaerobic degradation of *p*-cresol by an enriched bacterial consortium using a nonlinear model.

Indigenous microbes in secondary treated domestic wastewater did not require an acclimation period before degrading phenol, whereas those in landfill leachate exhibited a lag period (Deeley et al. 1985). Activated sludge followed by lagooning gave the best removals of phenol and pentachlorophenol in a comparison of six wastewater treatments (Hannah et al. 1986). Beltrame et al. (1984) developed a structure–toxicity relationship on the inhibiting effect of chloro- and nitrophenols on phenol degradation in activated sludge. In attempts to model the kinetics of biodegradation of phenol at concentrations of 1.17 and 2.35 ng/mL in sewage, Simkins et al. (1986) concluded that the poor fit of their chosen models was due to the sewage organisms, active in the degradation of phenol, utilizing both the added phenol as well as organic compounds in the sewage as substrates.

Rubin et al. (1982) suggested the existence of three types of microbes to account for the results they obtained for the rates of mineralization of phenol and other compounds in lake water and sewage. These were: (1) microbes that mineralize phenol at concentrations <1 µg/mL; (2) those that mineralize phenol at concentrations >1 µg/mL; and (3) those that metabolize phenol at low concentration but do not assimilate the carbon of phenol to any great extent. The method developed by Subba-Rao et al. (1982) enabled them to study the mineralization of phenol and several other aromatics in freshwater and sewage at very low concentrations. Mineralization of phenol in lakewaters was linear with time in the concentration range 102 fg to 10 mg/mL. At concentrations less than 300 ng/mL 96% of the phenol was mineralized with very little incorporation into cells. The authors suggested that a transformation similar to cometabolism may exist in those cases of mineralization without assimilation. They concluded that the kinetics of mineralization, the extent of assimilation of substrate carbon, and the sensitivity of the aquatic population is different at trace levels than at higher levels of organic substances. Naturally occurring nutrients in lakewaters appear to have a profound effect on the mineralization of trace amounts of pollutants such as phenol since rates of mineralization of phenol were related to the trophic level of the water (Rubin and Alexander 1983). The addition of arginine, yeast extract, or various inorganic nutrients often stimulated mineralization.

Rubin and Schmidt (1985) devised a method to enable them to enumerate both prokaryotic and eukaryotic phenol-mineralizing microbes in fresh water. Phenol-mineralizing rates were 6.3 times faster for the bacteria than the fungi at a phenol concentration of 200 pg/mL.

Models for the kinetics of biodegradation of organic compounds which are not supporting the growth of a microbial community have been developed by Schmidt et al. (1985). They studied the mineralization of phenol (1 ng/mL) by *Pseudomonas acidovorans* while the bacterium was growing on uncharacterized organic carbon in a synthetic medium and concluded that a first order model, or one that allows for the kinetics of growth of the population on other substrates, gave the best fit.

Chesney et al. (1985) concluded that no cometabolism of phenol takes place in freshwater at concentrations in the range 1 ng to 1 μg/mL. They found that about 80% of the phenolic carbon was mineralized and about 20% was assimilated and reasoned that phenol does not persist in fresh water because it is either mineralized or incorporated. Jones and Alexander (1986) found that usually < 10% of the phenolic carbon was incorporated into microbial cells metabolizing phenol at concentrations of 10 ng/mL or below. The figure was higher for higher concentrations of phenol. The mineralization of 2 μg/L of phenol by *Pseudomonas acidovorans* growing on uncharacterized organic substances in an inorganic salts solution was delayed when 70 μg/L of acetate was added, but when only 13 μg/L of acetate was added, both were utilized simultaneously (Schmidt and Alexander 1985). Second substrates and uncharacterized organic carbon appear to have a profound effect on the rate and extent of biodegradation of pollutants at trace concentration. Adaptation of a mixed aquatic population by long-term exposure to phenol resulted in adaptation to *m*-cresol, *m*-aminophenol, and *p*-chlorophenol (Shimp and Pfaender 1987).

3. Chlorophenols.   Vast amounts of chlorinated phenols are used each year and eventually find their way into water and wastewater where they constitute a major pollution problem. For example, Milner and Goulder (1986) detected 14 chlorophenols, nitrophenols, and phenoxyalkanoic acids in a stream in Yorkshire. Chlorinated phenols are also important products of biodegradation of many pesticides and can also be formed during pesticide degradation in soil. For example, Smith (1985a) detected 2,4-dichlorophenol in soil during the degradation of 2,4-D. Schwien and Schmidt (1982) studied a 3-chlorobenzoate-degrading *Pseudomonas* sp. which after adaptation was able to degrade 4-chlorophenol. Even after a 4-mon period of adaptation they were unable to induce an *Alcaligenes* sp. that could grow on benzoate and phenol, to utilize 4-chlorophenol as a sole source of carbon. Conjugation experiments with the two strains yielded transconjugants, *Alcaligenes* sp. strain A7-2, that could utilize 4-chlorophenol up to 5 m*M*, and even a mixture containing 2 m*M* of each of the three monochlorophenols. The transconjugants could also utilize catechol, 3-chloro-, 4-chloro-, and 3,5-dihlorocatechols and phenol.

The *Pseudomonas cepacia* AC1100 obtained by "molecular breeding" with the ability to utilize the herbicide 2,4,5-T as a sole carbon source has degraded a variety of halogenated phenols (Kilbane et al. 1982; Karns et al. 1983a and b). Among the various di- and tri-chlorophenols tested by Karns et al. (1983b), those

with two chlorine atoms in the one position on the aromatic ring, such as 2,3,6-
and 3,4,5-trichlorophenols, were degraded to a lesser extent than those with a
chlorine atom in each of the *o*-, *m*-, and *p*- positions, such as 2,4,5- and 2,3,4-
trichlorophenols.

Shimp and Pfaender (1985a and b) studied the effect of adaptation to naturally
occurring organic substrates by mixed microbial communities in lake waters on
the degradation of monosubstituted phenols. Humic acids were found to suppress
biodegradation of *p*-chlorophenol as well as cresol and *m*-aminophenol (Shimp
and Pfaender 1985b). However, adaptation to increasing concentrations of amino
acids, carbohydrates, or fatty acids enhanced the biodegradation of all three
phenols (Shimp and Pfaender 1985a). Extended adaptation of a mixed aquatic
microbial community to phenol was shown to lead also to adaptation of the community to degrade *p*-chlorophenol, *m*-aminophenol, and cresol (Shimp and
Pfaender 1987).

O-Methylation of mono-, di-, tri-, and tetrachloroguaiacols by two strains of
Arthrobacter-like bacteria (Neilson et al. 1983) and various fluoro-, chloro-, and
bromophenols by *Rhodococcus* sp. and *Acinetobacter* sp. (Allard et al. 1987) has
been reported. Allard et al. (1987) suggested that since *O*-methylation is highly
competitive with biodegradation pathways, the process may have environmental
significance because it produces lipophilic molecules that could be as toxic as the
parent compounds.

When *Pseudomonas* sp. strain B13 was present in a mixed microbial community in synthetic sewage containing phenol, acetone, and alkanols plus
4-chlorophenol or a mixture of chlorophenol isomers, the mixed substrates were
completely degraded (Schmidt et al. 1983). The *Pseudomonas* sp. played a key
role as the chlorocatechol-dissimilating organism. The authors detected the formation of the hybrid strain *Alcaligenes* sp. strain A7-2 which had acquired the
genes for chlorocatechol degradation from the *Pseudomonas* sp.

Since substituted phenols are known to undergo adsorption to soil (Boyd 1982)
they may also undergo sorption to sewage sludge.

Therefore the fate of these substances during anaerobic sludge digestion is
important. Boyd et al. (1983) studied the degradation of phenol and the three
isomers of chlorophenol, methoxyphenol, cresol, and nitrophenol in diluted
sewage sludge. All of the monosubstituted phenols with the exception of
p-chlorophenol and *o*-cresol were significantly degraded during an 8-wk period.
Complete mineralization of *o*-chlorophenol was found and the initial step in its
degradation appeared to involve dechlorination. In a later study on the degradation of mono- and dichlorophenol isomers by acclimated and unacclimated
sludge, Boyd and Shelton (1984) reported the degradation of all three monochlorophenols in unacclimated sludge, but the *p*-isomer was the most resistant.
Among the dichlorophenol isomers, 3-, 4-, and 3,5-dichlorophenol were persistent. Reductive dechlorination of the chlorine atom in the *o*-position with respect
to the phenolic group was found. These researchers detected specific cross-
acclimation patterns, for example, sludge acclimated to *p*-chlorophenol could

degrade all of the monochlorophenols as well as 2,4- and 3,4-dichlorophenol. With ^{14}C-labeled substrates, *o*- and *p*-chlorophenol and 2,4-dichlorophenol were degraded with the formation of $^{14}CO_2$ and $^{14}CH_4$. Reductive dechlorination of chlorophenols was also found in samples of anoxic aquifers, freshwater sediment and sewage sludge (Gibson and Suflita 1986). Sulfate inhibited dehalogenation in an aquifer sample from a sulfate-reducing site. This finding raises the question on how this might affect the degradation of halogenated compounds in marine sediments.

To find a suitable correlation to predict the biodegradability of organic compounds, Paris et al. (1982) compared the second-order rate constants for the transformations of a series of phenolic compounds by *Pseudomonas putida* with the physicochemical properties of the phenols. The relationship between microbial transformation rate constants and van der Waal's radii proved to be the best.

H. Anilines

Hybrid trimers were produced from 2,4-dichlorophenol and various halogenated anilines by combined chemical activity and the activity of a fungal phenol oxidase (Liu et al. 1981c). It was suggested that such hybridization of pesticide residues or pesticide residues and naturally occurring substances might also occur in soil.

Reduction of nitrate to nitrite in cultures of *Escherichia coli* in the presence of various anilines led to the chemical formation of diazo- derivatives of the anilines (Lammerding et al. 1982). Nitrite produced by *Escherichia coli* in the presence of 3,4-dichloroaniline resulted in the production of 3,4-dichlorophenol and a bis-(dichlorophenyl)triazene and small amounts of tetrachloroazobenzene, tetrachlorobiphenyls, and dihydroxytetrachlorobiphenyls.

A laccase of the white rot fungus *Trametes versicolor* polymerized 4-chloroaniline to produce oligomers ranging in size from dimers to pentamers (Hoff et al. 1985). The action of these extracellular laccases in the oxidative coupling of toxic intermediates of pesticide decomposition, if it occurs in soil, may be looked upon as a detoxication reaction.

The halogenated aniline, 3,4-dichloroaniline, a product of the biodegradation of a number of herbicides, such as the phenylamide herbicides, has generally been found to be resistant to mineralization. However, Zeyer and Kearney (1982b) mineralized ^{14}C-ring-labeled propanil in soils by inoculation of the soils with a *Pseudomonas* sp. that could utilize 4-chloroaniline as a sole source of carbon and nitrogen. The bacterium had been isolated previously from soils by chemostat culture (Zeyer and Kearney 1982a) and was also found to utilize aniline and 3-chloroaniline as sole carbon and nitrogen sources. Growth on 2-chloroaniline was much slower.

You and Bartha (1982a) employed analogue enrichment with propionanilide as the sole carbon source and isolated a strain of *Pseudomonas putida* that mineralized 3,4-dichloroaniline when propionanilide was present. Detailed studies with this bacterium revealed that the degradation of 3,4-dichloroaniline proceeded via 4,5-dichlorocatechol and 3,4-dichloromuconate. The authors partly attributed the slow mineralization of 3,4-dichloroaniline, usually found in soils, to reactions that reduce its availability to microorganisms such as polymerization and binding reactions. Oxidative coupling of substituted anilines and phenols by a horseradish peroxidase showed that the degree of binding to humic materials depended to a large extent on the type of aromatic ring substituent (Berry and Boyd 1984). The presence of a nitro- group rendered anilines and phenols nonreactive with the peroxidase. Extending the analogue enrichment method to soil, You and Bartha (1982b) found that the addition of aniline to soil that contained 3,4-dichloroaniline in both free and bound form led to greatly increased mineralization of the halogenated aniline. The authors favored specific enrichment of microbial populations and induction of enzyme systems that can cometabolize 3,4-dichloroaniline as explanations for the stimulation of mineralization. You and Bartha (1982b) felt that displacement of 3,4-dichloroaniline by aniline was less likely to be important. However, Saxena and Bartha (1983b) later showed that 3,4-dichloroaniline bound by humic acid was partially displaced by aniline and various substituted anilines. In view of the fact that Pillai et al. (1982) demonstrated the formation of nitrobenzene and p-benzoquinone from aniline in sterilized soil samples, the role of aniline in the stimulation of the mineralization of 3,4-dichloroaniline in nonsterile soils is still not clear. Since Saxena and Bartha (1983a) were also able to demonstrate that humic acid-(ring ^{14}C)-labeled 3,4-dichloroaniline complexes were mineralized at rates faster than soil organic matter, they expressed the opinion that an extensive accumulation of 3,4-dichloroaniline in soil humus is unlikely. Under anaerobic conditions in lake sediments and sludge samples, anilines were found to be persistent (Horowitz et al. 1982).

In pure culture studies with an isolate of *Pseudomonas putida*, You and Bartha (1982b) also demonstrated mineralization enhancement of 3,4-dichloroaniline in the presence of aniline. They stated that their approach may be useful in soil decontamination rather than the use of unique microbes.

Two bacteria isolated from soil for their ability to degrade aniline, *Rhodococcus* sp. and a pseudomonad, were unable to metabolize monochlorinated anilines unless they were supplied with additional carbon sources (Kaminski et al. 1983). Zeyer and Kearney (1983a) found that their soil isolate, a *Pseudomonas* sp. grew slowly on p-nitroaniline as a source of carbon but mineralized the ring carbon much more rapidly in the presence of yeast extract. The bacterium could also mineralize m-nitroaniline, but it required the presence of p-nitroaniline in the medium. Anson and Mackinnon (1984) isolated a *Pseudomonas* sp. capable of growth on aniline as a sole carbon source. Aniline metabolism by the bacterium was associated with a plasmid. A strain of *Moraxella* sp. able to use aniline and a number of monohalogenated anilines, including 4-chloroaniline, as sole sources of carbon and nitrogen, was found to convert 4-chloroaniline to 4-chlorocatechol followed by an ortho- cleavage of the aromatic ring (Zeyer et al. 1985). The aniline oxygenase involved in the first step was inducible by a number of aniline compounds and showed broad substrate specificity.

Biodegradation was the most significant means of removal of aniline in pond water and its mineralization was accelerated when the samples were inoculated with sewage sludge (Lyons et al. 1984). The major pathway of aniline degradation involved oxidative deamination to catechol followed by ring fission. Aniline was mineralized by a *Pseudomonas* sp. when it was supplied at 3 ng/mL in a mineral salts solution but only after growth on uncharacterized soluble organic carbon had ceased (Schmidt and Alexander 1985). When supplied along with glucose at 300 ng/mL aniline was not mineralized until the glucose had largely been converted to carbon dioxide. The authors suggested that second substrates of either known or unknown chemical character may play an important role in governing the rate and extent of biodegradation of low concentrations of substances such as aniline in natural environments. Inoculum composition and size was found by Kool (1984) to have a substantial effect on the success of biodegradability tests using 4-nitrophenol and three methoxyaniline isomers as test compounds. Higher inoculum rates gave better reproducibility. Aniline mineralization rates in lakewater were proportional to concentration in the range 100 pg/mL to 1 µg/mL (Hoover et al. 1986). Below 100 pg/mL, however, the rates were less than the linear relationships indicated, and this again raises the question of what concentration of pollutants should be used in biodegradation tests.

Photochemical processes in water samples from an eutrophic lake accounted for most of the mineralization of 2,4,5-trichloroaniline, a metabolite in the degradation of a number of phenylurea and phenylcarbamate herbicides (Hwang et al. 1985). Only abut 19% of the ring-^{14}C-labeled 2,4,5-trichloroaniline was mineralized by biological processes and this was attributed to bacterial activity.

Paris and Wolfe (1987) obtained a good correlation between microbial transformation rate constants and van der Waal's radii in their bacterial transformations of aniline and a series of monosubstituted anilines. A chlorine substituent reduced the transformation rate by a factor of 10 for a pure culture of a gram-negative rod-shaped bacterium isolated from river water. Larger substituents such as a methyl group decreased the rate of transformation. Although rates were higher for a concentrated natural population from pond and river waters, the relationships of substituent size and transformation rates were similar.

I. Amides

Low concentrations of the amide herbicides alachlor and propachlor were not mineralized in tests with sewage and lakewater; rather they were cometabolized to organic products (Novick and Alexander 1985). Although aniline represents the structural nucleus of these herbicides, the addition of aniline to sewage did not increase the formation of cometabolites from propachlor. Instead, glucose addition brought about an increase in the products of its cometabolism. Cometabolism of these herbicides appears to be important in their transformation at low concentration in natural waters.

Propachlor was found to undergo mineralization to a considerable extent in soil suspensions of a soil that had been treated previously in the field with the herbicide (Novick et al. 1986). Since the ^{14}C-ring-labeled alachlor used in the study was only 97% pure the authors suggested that the small amount of $^{14}CO_2$ evolved may have come from contaminants rather than the herbicide. Although mineralization of alachlor was very low or nonexistent and attempts to isolate microbes capable of mineralizing it were unsuccessful, the herbicide appeared to undergo cometabolism in both soil suspensions and the aquifer samples as organic products were detected. Very little of the ^{14}C-ring-labeled alachlor was mineralized. A mixed culture of two bacteria was isolated that could bring about extensive mineralization of propachlor, producing metabolite identified as *N*-isopropylaniline.

Bollag et al. (1986) concluded that biological processes were responsible for the reductive dechlorination of alachlor under anaerobic conditions in a stream sediment. They identified 2′,6′-diethyl-*N*-(methoxymethyl)acetanilide as the metabolite. The authors did not comment on the mineralization of alachlor in their experiments, but presumably it was low or nonexistent as their recovery levels of total ^{14}C label indicated no significant formation of other products.

Although efforts to isolate microorganisms that could use propachlor as a sole source of carbon and energy failed, several bacteria were isolated from sewage and lakewater that could (Novick and Alexander 1985). Two of the bacterial isolates metabolized the herbicide without assimilating any of the molecular carbon. Cometabolism of propachlor appears to be an important reaction governing the fate of the pesticide in natural waters when present at low concentration. However, mineralization of propachlor was detected when it was added to soil suspensions at concentrations of 25 ng and 10 µg/mL (Novick et al. 1986). These workers isolated two bacteria, a gram-negative rod and a gram-positive coccus, that in mixed culture could mineralize propachlor. Neither bacterium could mineralize the ring carbon when grown as a pure culture. *N*-Isopropylaniline was identified as a product of the bacterial metabolism of propachlor. They concluded that mineralization of propachlor was an important factor governing its fate in soil.

The contact herbicide propanil was degraded in a soil amended with ring-^{14}C-labeled propanil and an inoculum of a *Pseudomonas* sp. that could utilize 4-chloroaniline as a sole source of carbon and nitrogen (Zeyer and Kearney 1982b). The enzymes responsible for the degradation of the aromatic ring were inducible by 4-chloroaniline and 3,4-dichloroaniline. Wright and Maule (1982) found that two species of green algae, *Chlamydomonas reinhardii*, *Chlorella vulgaris*, and four blue-green bacteria, *Anabaena cylindrica*, *Anacystis nidulans*, *Gloeocapsa alpicola*, and *Tolypothrix tenuis*, could hydrolyze propanil with the formation of 3,4-dichloroaniline.

Anaerobic degradation of propanil in an enrichment inoculated with a pond sediment led to the formation of possibly 3-chloropropioanilide and an unknown compound (Stepp et al. 1985). Soil enrichment cultures containing propanil plus yeast extract and tryptone, held under anaerobic conditions, led to the formation of 3,4-dichloroaniline (Pettigrew et al. 1985). Secondary enrichment cultures developed from an inoculum of the supernatant of the primary culture produced 3,4-dichloroaniline and *m*-chloroaniline from propanil.

J. Urea Pesticides

The fate of the two urea herbicides, linuron and monuron, in soil has been reviewed by Maier-Bode and Hartel (1981). A study of the anaerobic degradation of diuron, using enrichment cultures developed from a pond sediment inoculum, showed that it was quite different from aerobic degradation (Attaway et al. 1982). Degradation of diuron in enrichment cultures was completed in 17 to 25 d. A major product was 3-(3-chlorophenyl)-1,1-dimethylurea. Later work in the same laboratory with anaerobic enrichment cultures revealed that linuron was also dehalogenated at the *p*-position (Stepp et al. 1985). This biological reaction yielded 3-chlorophenyl-1-methoxy-1-methylurea.

The urea insect growth regulator diflubenzuron, which interferes with the deposition of insect chitin, was found to be mineralized in a number of agricultural soils (Nimmo et al. 1984). The degradation proceeded via the hydrolytic cleavage of the molecule to 4-chlorophenylurea and 2,6-difluorobenzoic acid, after which both phenyl rings were cleaved. Nimmo et al. (1986) studied the fate of ^{14}C-labeled 4-chlorophenylurea labeled on the phenyl ring or the carbonyl group. They detected mineralization of ^{14}C from both sites indicating its extensive microbial degradation in soils. Chapman et al. (1985) also obtained evidence suggesting that microbial activity is important in diflubenzuron degradation in soil.

K. Metallo- and Metallo-organic Compounds

Bis(tributyltin)oxide shows a very broad spectrum of activity, and has been used as a fungicide, bactericide, algicide, and molluscicide. Its chief uses are in marine antifouling paint formulations, as an algicide in cooling tower waters, and as a wood preservative. While unsuccessful in attempts to isolate microbes able to use bis(tributyltin)oxide as sole carbon source, Barug (1981) found that cultures of *Pseudomonas aeruginosa* and *Alcaligenes faecalis* and the fungi *Coniophora puteana*, *Trametes versicolor*, and *Chaetomium globosum* could degrade the product when the organisms were growing in a nutrient-rich medium. The chief metabolite was monobutyl tin.

Dilute solutions of the widely used wood preservative, chromated copper arsenate, and wood soaked in the preservative are subject to attack by the fungus *Candida humicola* with the production of trimethylarsine (Cullen et al. 1984).

Both chemical and biological factors appear to be involved in the degradation of the fungicide mancozeb in soil (Doneche et al. 1983). Enrichment cultures yielded a culture of *Bacillus* sp. that degraded the fungicide, which is a coordination product of zinc ion and manganous ethylene-bis-dithiocarbamate.

L. Pyrethroids

Synthetic pyrethroid insecticides have shown promise as replacements for the organophosphates. Microbial degradation appears to be an important factor in the persistence of pyrethroids in soil. Chapman et al. (1981) found "heat labile factors" played a major role in the degradation of fenpropanate, permethrin, cypermethrin, fenvalerate, and decamethrin.

Permethrin degradation was enhanced in a silt loam amended with either sewage sludge or dairy manure (Doyle et al. 1981). Fenvalerate residues in sugar cane trash applied to soil also appeared to be subject to microbial degradation (Smith and Willis 1985). The persistence of fenvalerate and permethrin in the estuarine environment was relatively short with half-lives of 34 d and <2.5 d, respectively (Schimmel et al. 1983). Fluvalinate disappeared rapidly from three soil types under aerobic and anaerobic conditions (Staiger and Quistad 1983). Although no sterile soil controls were used in the study, biodegradation appeared to play a major role in fluvalinate degradation in the sandy loam, clay loam, and clay soils studied.

M. Miscellaneous Pesticides and Related Compounds

A metabolite of the acaricide and insecticide carbophenothion, *p*-chlorophenyl-methyl sulfide and the corresponding sulfoxide and sulfone were found to undergo mineralization in a sandy loam (Guenzi and Beard 1981). Mineralization of the compounds was correlated with soil organic carbon mineralization.

Benzazimide is one of the important degradation products formed in soil from the phosphorodithioate insecticide and acaricide azinphos-methyl. Engelhardt and Wallnöfer (1983) have tested a wide range of bacteria for their ability to degrade benzazimide. Of the 33 strains tested, representatives from *Corynebacterium, Brevibacterium, Arthrobacter, Nocardia*, and *Pseudomonas*, they found that only one strain of *Pseudomonas* sp. was able to transform benzazimide, and that to 5-hydroxybenzazimide.

The fungi, *Botrytis cinerea, Coriolus versicolor, Cladosporium cucumerinum*, and *Fusarium culmorum*, metabolize the systemic triazoline fungicide, triadimefon, to triadimenol (Deas et al. 1984a and 1984b). Different amounts and types of triadimenol enantiomers were produced.

Warfarin, an anticoagulant rodenticide, is the racemate of 4-hydroxy-3-(3-oxo-1-phenylbutyl)coumarin. *Nocardia corallina* exhibits stereoselective reduction of *S*-warfarin to the *S*-warfarin-*S*-alcohol (Davis and Rizzo 1982). An *Arthrobacter* sp. proved capable of the reduction of both members of the racemate to the corresponding *S*-alcohols.

Inoculation of different soil types with an acclimated culture of *Pseudomonas aeruginosa* revealed that it degrades TMTD (Tetramethylthioperoxydicarbonic diamide) in a wide variety of soils and could tolerate a concentration of the fungicide as high as 500 µg/g (Shirkot and Gupta 1985).

Sariaslani and Rosazza (1985) isolated and characterized three major products from the transformation of the botanical rotenoid insecticide, 1',2'-dihydrorotenone, by *Streptomyces griseus* and these were 1',2'-dihydro-6aβ-hydroxyrotenone, 1',2'-dihydro-2',6aβ-dihydroxyrotenone and 1',2'-dihydro-1,6aβ-dihydroxyrotenone.

The green alga, *Chlorella* sp., and a blue-green bacterium, *Oscillatoria* sp., degraded the formamidine acaricide chlordimeform with the production of $^{14}CO_2$ from ring-labeled chlordimeform (Benezet and Knowles 1981). The initial reaction, an hydrolysis, was suspected to be nonenzymatic and yielded 4-chloro-*o*-formotoluidine.

Enrichment cultures of soil with the widely used nitrile herbicide bromoxynil, yielded several bacterial isolates capable of metabolizing the herbicide (McBride et al. 1986). *Klebsiella pneumoniae* subsp. *ozaenae*, one of the isolates obtained from the enrichment cultures, converted bromoxynil(3,5-dibromo-4-hydroxybenzonitrile) to 3,5-dibromo-4-hydroxy benzoic acid. The ammonia liberated during this conversion was used as a sole nitrogen source by the bacterium. Nitrilase activity was detected in cell-free preparations of the bacterium and this enzyme activity proved to be highly specific for bromoxynil.

The fungicide pentachloronitrobenzene is metabolized by the protozoan *Tetrahymena thermophila* with the formation of nitrite, pentachlorothioanisole, and pentachloroaniline (Murphy et al. 1982). Evidence indicated that the pentachlorothioanisole may have been formed via a glutathione-dependent pathway.

Biodegradation of the pyridilium herbicide paraquat by a soil yeast, *Lipomyces starkeyi*, occurred when used in the culture medium as the sole source of nitrogen (Carr et al. 1985). Using ^{14}C-labeled pesticide, the formation of both $^{14}CO_2$ and ^{14}C-acetic acid was detected.

Although Saxena et al. (1987) were unsuccessful in isolating microorganisms that could mineralize the germination inhibitor metolachlor, they demonstrated that it could be transformed by some culture collection strains of *Bacillus circulans*, *Bacillus megaterium*, *Fusarium* sp., *Mucor racemosus*, and an actinomycete isolated previously by Krause et al. (1985). The chief metabolites produced by the actinomycete were shown by nuclear magnetic resonance and mass spectrometry to be 2-chloro-*N*-(2-ethyl-6-hydroxymethylphenyl)-*N*-(2-methoxy-1-methylethyl) acetamide and the demethylation product 2-chloro-*N*-(2-ethyl-6-methylphenyl)-*N*-(2-hydroxy-1-methylethyl)acetamide.

Stralka and Camper (1981) considered that cometabolism appeared to be involved in the microbial degradation of profluralin as they could detect degradation of the herbicide by their isolates only when the cultures were provided with additional sources of carbon and nitrogen. They tentatively identified the bacteria as strains of *Arthrobacter simplex*, *Cellulomonas flavigenum*, and *Micro-*

bacterium flavum. The metabolites produced by *M. flavum* were found to be 2-amino-α-α-α-trifluoro-6-nitro-*N*-propyl-*p*-toluidine, *N*-(cyclopropylmethyl)-α-α-α-trifluoro-2,6-dinitro-*p*-toluidine, and 2-amino-α-α-α-trifluoro-6-nitro-*p*-toluidine.

About 60 of the soil isolates obtained by Zeyer and Kearney (1983b) mineralized the carbon of the propyl-1-^{14}C group of trifluralin to some extent, indicating that the ability to dealkylate trifluralin is widespread among soil microbes. A yeast, *Candida* sp., which mineralized the propyl-1-^{14}C at the highest rate (11%) was unable to mineralize the ring-^{14}C. A metabolite, α-α-α-trifluoro-2,6-dinitro-*p*-toluidine, was tentatively identified as one of the products of metabolism.

The herbicide fenac (2,3,6-trichlorophenylacetic acid) is persistent in soil and was also only slowly degraded in some aquatic ecosystems such as lakewater and sewage water (Rosenberg 1984). However, Rosenberg found that uniformly ring^{14}C-labeled fenac was mineralized in a temperate soil of pH 6.9 and organic matter content of 2.1%. A product of the degradation of fenac formed during incubation in nutrient broth-supplemented sewage was tentatively identified as a toluene having two chlorine atoms and an hydroxy group as substituents on the aromatic ring.

Endothal, a pre- and post-emergence herbicide, was rapidly mineralized in a shake flask test using reservoir water and was considered environmentally safe for use in the control of aquatic weeds (Reinert et al. 1986).

Cometabolism of 4-chloro-3,5-dinitrobenzoic acid by a nonaxenic culture of *Chlamydomonas* sp. resulted in the formation of α-hydroxymuconic semialdehyde which could be mineralized by a strain of *Streptomyces* sp., demonstrating that combined cometabolism and metabolism can lead to its degradation (Jacobson and Alexander 1981). Johnson and Williams (1982) found that the addition of sodium benzoate as growth substrate to a culture of *Pseudomonas fluorescens* enabled cometabolism of 3-chlorobenzoate to occur with the production of 3-chlorocatechol. When the chlorocatechol accumulated, the 3-chlorobenzoate could be metabolized by ring fission and dehalogenation.

The cometabolism of chlorobenzoic acids by a *Pseudomonas* sp. was studied by Veerkamp et al. (1983), who found that only 3-chlorobenzoic acid was cometabolized in batch culture, whereas all three chlorobenzoate isomers were cometabolized in continuous culture. They found that only 4-chlorocatechol induced catechol 2,3-dioxygenase activity.

Anaerobic degradation of chloro-, bromo-, and iodobenzoates by lake sediment was demonstrated by Horowitz et al. (1983). The degradation proceeded via reductive dehalogenation. Studies on the kinetics of dehalogenation of 3-chlorobenzoate, 3,5-dichlorobenzoate, and 4-amino-3,5-dichlorobenzoate in anoxic sediment slurries revealed that the initial dehalogenation reaction followed Michaelis–Menten kinetics (Suflita et al. 1983).

Marks et al. (1984) isolated an *Arthrobacter* sp. from sewage sludge that utilized 4-chlorobenzoate as a sole source of carbon. The degradation involved an initial dehalogenation to give 4-hydroxybenzoate which was metabolized via protocatechuate. A strain of *Pseudomonas alcaligenes* isolated from activated sludge utilized 3-chlorobenzoate as a sole carbon source (Focht and Shelton 1987). Inoculation of 3-chlorobenzoate-amended soil with the bacterium led to complete degradation of the 3-chlorobenzoate.

Alcaligenes denitrificans NTB-1, able to use 4-chlorobenzoate, could also use 4-bromo-, 4-iodo-, and 2,4-dichlorobenzoate as sole sources of carbon (van den Tweel et al. 1987). Metabolism of the monohalobenzoates was initially by hydrolytic dehalogenation to 4-hydroxybenzoate whereas the dichlorobenzoate was reductively dehalogenated to yield 4-chlorobenzoate.

Pignatello (1986) demonstrated that ethylene dibromide was mineralized by aerobic microbes present in a stream bottom sand of low organic carbon and in a muddy soil of high organic carbon. Only a few days was necessary for the degradation of ethylene dibromide to reduce the concentration from 6 to 8 ng/mL to near or below the limit of detection (20 pg/mL).

Enhanced degradation of the fungicides iprodione and vinclozolin have been demonstrated in a sandy loam soil and the enhanced activity could be transferred to a less active soil by a 10% amendment with an active soil (Walker et al. 1986).

The bactericide techlofthalam [N-(2,3-dichlorophenyl)-3,4,5,6-tetrachlorophthalamic acid] has been used to control bacterial leaf blight in rice. Reductive dechlorination of the tetrachlorophthalamic acid moiety occurs in relatively shallow layers of flooded rice soil (Kirkpatrick et al. 1981), and some mineralization of the [14]C.

IV. Bound Pesticide Residues

Bound pesticide residues appear to form as a result of chemical (Stevenson 1976), enzyme-catalyzed (Sjoblad and Bollag 1981) or physical (Khan 1980 and 1982a and b) interactions or combinations of these three with pesticides or their degradation products and the humic acid fractions of soils. Khan (1982b) proposed that bound residues may be defined as "chemical species originating from pesticide usage, that cannot be extracted by methods commonly used in residue analysis," provided that the chemical species referred to are "either the intact pesticide or compounds derived from it." He stressed that the extraction procedure must be specified.

Lichtenstein (1980) has reemphasized the need for a more thorough understanding of the nature of bound residues in soils particularly in relation to their bioavailability to crops and soil organisms. Khan (1982b) has also stressed the importance of bioavailability data and has proposed working definitions for bioavailable and biounavailable bound residues.

Numerous studies with ^{14}C-labeled pesticides and their transformation products have revealed the formation of bound residues in soils and sediments. Khan

(1980 and 1982a and b) has listed some of these, and some others published since then are given in Table 2.

Bound residues were detected in soil samples by combustion of the residual ^{14}C to $^{14}CO_2$ after exhaustive extraction with nonpolar organic and polar solvents. However, very little knowledge of the chemical nature of the bound residues is available. The work of Khan and his associates has shed some light on this problem. Khan and Hamilton (1980) devised a high temperature (800°C) distillation procedure that enabled them to distill and trap bound residues from a prometryn-treated organic soil. GC-MS of distillate fractions revealed that a considerable portion of the bound residue in the organic soil was in the form of prometryn. The bound ^{14}C residues were associated with the humin, humic acid, and fulvic acid fractions of the soil (Khan 1980) and were found to consist largely of prometryn and 2-(methylthio)-4-amino-6-(isopropylamino)-s-triazine, its monodealkylated product. The finding of unextractable residues of prometryn, which was only made possible by the high temperature distillation of the extracted soil, led Khan (1982a and c) to propose another mode of formation of bound residues—one of physical entrapment of the residues in the lattice-like structure formed by humic materials. Conventional solvent extraction fails to remove these substances. Khan has drawn the analogy of a molecular sieve function of such a structure. High-temperature distillation apparently leads to thermal weakening of the lattice structure, enabling the distillation and removal of the residues.

Acknowledging the fact that the high-temperature distillation method leads to considerable decomposition of the bound residues to CO_2, thus giving low recoveries of the residues and therefore making chemical characterization more difficult, Capriel et al. (1986) compared their high temperature distillation method with an extraction procedure using supercritical methanol. Supercritical fluids which can be formed at high temperature and pressure are considered to be forms of matter that are distinctly different from solids, gases, and liquids. They are often considered to be intermediate between liquids and gases from which they are distinguished by density (Josephson 1982). Capriel et al. (1986) used a temperature of 250°C and a high-performance liquid chromatography (HPLC) pump to generate the high pressure and temperature needed to convert methanol, the chosen solvent, into the supercritical state. The HPLC column contained the soil sample to be extracted. With this technique they obtained results that were superior to those obtained by the high-temperature distillation technique. Their study was concerned with the formation of bound residues of atrazine, prometryn, deltamethrin, diuron, 2,4-D, and methyl parathion in a variety of soils from North America and Germany.

Many pesticides and their degradation products have chemical structures similar to those of naturally occurring humus constituents and may become covalently linked to soil humus during the humification process. Bound in this way, by either enzymatic or chemical oxidative coupling reactions (Sjoblad and Bollag

Table 2. Bound residues of some pesticides and pesticide transformation products.

Pesticide or Product	Site of ^{14}C label	Time	Environment	Bound %	References
Alachlor	Ring	10 wk	Anaerobic stream sediment	50	Bollag et al. (1986)
Carbaryl	Naphthyl-1-^{14}C	40 d	Flooded laterite soil	18–22	Rajagopal and Sethunathan (1984)
			Flooded alluvial soil	20–22	
Carbofuran	Phenyl		Flooded laterite soil	25	
			Flooded alluvial soil	29	
Chlorpyrifos	Ring	32 d	Silt loam soil	15.6	Getzin (1981)
			Clay loam soil	12.8	
p,p′-DDT	Phenyl	13 d	Loam soil	2.8	Fuhremann and Lichtenstein (1980)
			Sandy soil	0.7	
3,4-dichloroaniline	Ring	150 d	Sandy loam	83.5	Saxena and Bartha (1983b)
2,6-dichlorobenzonitrile	Nitrile	61 d	Sandy humus soil	10.6	Chowdhury et al. (1981)
Diflubenzuron	Chlorophenyl	32 d	Sandy clay soil	approx. 57	Nimmo et al. (1984)
Fenitrothion	Ring	65 d	Black earth soil	73.7	Mac Rae (1986b)
			Red yellow podzol	59.4	
Fonofos	Phenyl	13 d	Loam soil	15.6	Fuhremann and Lichtenstein (1980)
			Sandy soil	9.3	
		14 d	Loam soil	23.7–28.3	Lichtenstein et al. (1982)
		28 d		30.8	Lichtenstein et al. (1983)
Lindane	Ring	13 d	Loam soil	4.6	Fuhremann and Lichtenstein (1980)
			Sandy soil	1.3	

Table 2. (*Continued*)

Pesticide or Product	Site of ^{14}C label	Time	Environment	Bound %	References
Methyl parathion	Ring	28 d	Sandy loam	26–33	Ou et al. (1983)
			Silty clay loam	35–45	
		49 d	Silt loam	47	Gerstl and Helling (1985)
Parathion	Phenyl	13 d	Loam soil	29.1	Fuhremann and Lichtenstein (1980)
			Sandy soil	6.0	
		14 d	Loam soil	31.7	Lichtenstein et al. (1982)
		28 d		48.5	
		28 d		43.8	Lichtenstein et al. (1983)
Pentachlorophenol	Ring	136 d	Flooded sandy soil	28.6	Weiss et al. (1982)
Phorate	Methylene	13 d	Loam soil	13.1	Fuhremann and Lichtenstein (1980)
			Sandy soil	4.5	
Prometryn	Ring	150 d	Humic mesisol	43	Khan and Hamilton (1980)

1981, Bollag and Loll 1983), they are nonextractable. Phenolic compounds are of particular importance in these reactions because phenols are major structural components of humic material. Kassim et al. (1981) found that 74% of the ring carbons of ^{14}C-catechol were stabilized into soil humus in a sandy loam over a 12-wk period. Oxidative coupling reactions catalyzed by a phenol oxidase from a soil fungus, *Rhizoctonia praticola*, have been shown to lead to the formation of a variety of phenolic and quinonoid oligomers from 2,4-dichlorophenol (Minard et al. 1981). Incubation of 2,4-dichlorophenol and various halogenated anilines with a fungal phenol oxidase, a scenario that could exist with mixed pesticide applications to soil, led to the formation of hybrid oligomers (Liu et al. 1981c). Both enzymatic and chemical reactions were involved in the formation of the hybrid molecules. In an attempt to show that these oxidative coupling reactions were not just an in vitro phenomenon and could occur in soil, Suflita and Bollag (1981) studied the polymerization of guaiacol, 1-naphthol, and 4-chloro-1-naphthol by an enzyme complex extracted from soil. They detected oligomer formation by coupling reactions and concluded that these results gave further support to the idea that oxidative coupling reactions are important in soil and in pesticide transformations and interactions between pesticides and humus.

Other workers have directed their attention to the effects of various soil treatments and amendments on the formation of bound residues. Amendments such as cow manure, sewage sludge, and the herbicide atrazine led to an increase in bound residues of fonofos whereas ammonium sulfate and the fungicide captafol had no effect (Lichtenstein et al. 1982). The formation of bound residues of parathion was stimulated through additions of manure and sludge. Additions of ammonium sulfate, urea, or K_2SO_4 to a flooded alluvial and a flooded laterite soil caused a decrease in bound residues of carbaryl and carbofuran (Rajagopal and Sethunathan 1984). Bound residue formation was more rapid in moist than in dry soil in the case of methyl parathion (Ou et al. 1983) and was very pronounced in the case of fonofos but particularly so for parathion in flooded soil (Lichtenstein et al. 1983). Mixing fonofos and parathion with soil also led to increased bound residues. The very high level of parathion-bound residues (71%) under flooded conditions was attributed by Lichtenstein et al. (1983) to the fact that the low redox potential of the soil would favor the metabolism of parathion to amino-derivatives.

Opinions are divided on whether bound pesticide residues pose an environmental problem and this question seems far from being answered. Release of bound prometryn from soil by microbial activity (Khan and Ivarson 1981) and from a soil suspension by UV irradiation (Khan 1982c) has been reported. Anhydrous ammonia applications to soil led to the displacement of bound 3,4-dichloroaniline and was considered a possible means of crop contamination (Saxena and Bartha 1983b). On the other hand, a wide range of inorganic and organic amendments to a silt loam soil failed to release detectable amounts of bound residues of methyl parathion. The amendments did however lead to the mineralization of the bound residues to $^{14}CO_2$ (Gerstl and Helling 1985). Organic amendments in the form of plant residues as well as unlabeled fenitrothion and 3-methyl-4-nitrophenol to a black earth and a red yellow podzolic soil were also found to stimulate the mineralization of bound residues of fenitrothion in the two soils (MacRae 1986b). You and Bartha (1982a) found enhanced degradation of bound 3,4-dichloroaniline through the addition of aniline to soil. This was attributed to analogue enrichment of the soil population and enzyme induction rather than displacement of the 3,4-dichloroaniline from humus.

Stott et al. (1983) expressed the opinion that chlorocatechols, degradation products from a wide range of pesticides having structures based on chlorinated aromatic rings, are not likely to be a source of future contamination once they have been linked into soil humic acid polymers.

Another interesting approach in studies of the release of substituted phenol bound residues has been the use of veratrylglycerol-β-aryl ethers as model compounds for soil-bound residues of these phenols. Methylthio-phenols, metabolites of a number of insecticides, formed model β-arylethers of veratrylglycerol that proved to be very resistant to degradation by cultures of *Rhodococcus equi* and *Corynebacterium equi* (Wallñofer et al. 1981). No cleavage of the ether linkage was detected and this may indicate that the methylthio-phenols are strongly bound to soil humus. On the other hand, β-aryl ethers formed from veratrylglycerol having either phenyl-,4-methoxyphenyl-, or 2,4-dichlorophenyl- substituents were readily cleaved between the C2 and C3 of the glycerol moiety by cultures of *Arthrobacter, Brevibacterium, Corynebacterium, Nocardia,* and *Rhodococcus* yielding phenoxyacetic acid, 3,4-dimethoxybenzoic acid and 2,4-dichlorophenoxy acetic acid (Engelhardt et al. 1981). If this type of microbial release of bound substituted phenols occurs in soil, it could be important from an ecotoxicological view.

V. Removal of Pesticides from Soil and Water

While the exploitation of biodegradation and other forms of interaction between microbes and pollutants has great potential in the removal of pesticide residues from soil and water and in the disposal of hazardous waste chemicals, much is left to be done if the full potential of biodegradation is to be realized. Some aspects of the problems involved have been discussed (Kobayashi and Rittman 1982; Speece 1983; Josephson 1984a; Dean-Ross 1987).

Experimentally, the removal of pesticides from water and waste water has been studied using various approaches including immobilized cells or enzymes, fixed microbial films, soil columns, and conventional wastewater processes. In the case of soil, inoculation with natural isolates or constructed bacterial strains and stimulation of indigenous microbial populations have been the major approaches to removal of pesticides. Sims et al. (1986) have compiled a comprehensive list of methods for use in surface soil decontamination.

Advances in our knowledge of cometabolism have opened up possibilities for the utilization of the process in removing pesticides from soil and water (Dalton and Stirling 1982; Janke and Fritsche 1985). Equally challenging are the possibilities flowing on from the rapid developments in genetic engineering which may allow the exploitation of constructed microbial strains in the degradation and/or detoxication of pesticides. Such aspects have been discussed by Pemberton (1981), Pemberton and Wynne (1984), Kilbane (1986), and Omenn (1986).

Pore and Sorenson (1981) found that cells of the alga *Prototheca zopfii* immobilized in agar beads were as efficient as activated charcoal in removing the insecticide kepone from water. Partitioning of the hydrophobic chlorinated hydrocarbon oxy- derivative was only one of the factors involved as living algal cells were more effective than dead algal cells. The pesticide was not degraded by the alga.

Fixed films of bacterial cells on a variety of supports have been found to be successful in removing a wide range of pesticides and structurally related substances from water and wastewater. Bouwer and McCarty (1982) reported the removal of chlorinated benzenes and aliphatics by a granular activated carbon column (GAC) with biological activity over a 2-yr period at an efficiency level of 95 to 98%. The combination of adsorption and biodegradation by the biologically active GAC led to the high level of removal efficiency.

Moore et al. (1985) developed a sand filtration plus carbon treatment system that was capable of removing 79% pyrethroids, 92% organophosphates, and 96% organochlorines from contaminated sea water. DeLaat et al. (1985) found that the role of the bacterial population on GAC columns in the removal of a selection of biodegradable substances from water came into play after the period of acclimation before which adsorption was most important.

A preparation of natural magnetite, consisting of microscopic particles of magnetic iron oxide, is used in water clarification known as the Sirofloc process (Austep Pty. Ltd. registered trademark) for the removal of color and turbidity components. This preparation was also found to adsorb large amounts of microbial cells (MacRae and Evans 1983 and 1984). After selecting bacteria for their ability either to metabolize or accumulate pesticides in their cells, it was shown that fixed films of these selected bacteria on the magnetic iron oxide particles could remove significant amounts of pesticides from water and waste water. During a contact period of 1 h, a fixed film of a mixture of *Rhodopseudomonas sphaeroides* (the accumulator) and *Alcaligenes eutrophus* (the degrader) removed 76.4% of 2,4-D and 33% of the lindane in water (MacRae 1985). A fixed film of *Rhodopseudomonas sphaeroides* on the magnetite removed 32.1% alpha hexachlorocyclohexane, 30% γ-hexachlorocyclohexane, 89.9% dieldrin, 90.8% heptachlor, 92.9% aldrin, and 100% DDT from a spiked wastewater in 20 min (MacRae 1986a).

The enzymatic hydrolysis of concentrated diazinon up to 1% in a sandy loam was demonstrated by Barik and Munnecke (1982) after a preparation of parathion hydrolase from a *Pseudomonas* sp. was added to the soil. The authors suggested that it could be used to clean up spills of concentrated organophosphorus pesticides. Johnson and Talbot (1983) discussed the practical use of microbial enzymes to detoxify pesticides in environmental clean-up procedures.

Soil inoculation with selected bacteria to detoxify particular pesticides has met with some success. For example, Edgehill and Finn (1983) have shown increased disappearance of pentachlorophenol in soil following inoculation with a PCP-degrading *Arthrobacter* sp. and Kilbane et al. (1983) were successful in removing 2,4,5-T from heavily contaminated soil by inoculation of the soil with their 2,4,5-T-degrading *Pseudomonas cepacia.*

In situ degradation of soil and water pollutants by introduced microbes has been a tantalizing goal for many microbiologists over the last thirty years. The rapid development in fairly recent times of plasmid technology and cloning techniques has brought this goal even closer. Ghosal et al. (1985) discussed the significance of genetic studies in the removal of toxic chemicals from the environment and suggested that the construction of appropriate degradative strains will be a key component in the solution of toxic chemical pollution problems. However, the successful application of these constructed strains to practical problems of soil and water contamination still requires solutions to problems of an ecological nature such as the survival, and establishment of the constructed strain in the soil, water, or wastewater environment. Finn (1983) expressed the view that highly specialized genetically engineered strains are unlikely to perform well in activated sludge units when the wastewater contains considerable amounts of other easily available growth substrates. Haas (1983) appears skeptical of the use of constructed strains for the solution of practical problems of biodegradation. Stotzky and Babich (1986) have drawn attention to the great lack of research on the effects of both abiotic and biotic environmental factors on such things as survival, establishment, and growth of, and genetic transfer by genetically engineered bacteria released into the environment and favor caution in their use. With very few exceptions inoculation studies fail through lack of knowledge as to how the introduced strain will survive and compete with the indigenous population in soil, water and wastewater.

Goldstein et al. (1985) investigated some of the reasons why inoculation of soil and water fail to remove the pollutant. *Pseudomonas* strains isolated from soil for the ability to mineralize 2,4-dichlorophenol and *p*-nitrophenol failed to mineralize 2,4-dichlorophenol in lake water, sewage and soil. Mineralization was detected when the bacterium was inoculated into sterile soil and to a lesser extent when the bacterium was inoculated into sterilized sewage. They suggested the reasons for failure to achieve degradation of pollutants in natural environments after inoculation might include: "(i) concentration of the pesticide in nature may be too low to support growth, (ii) the natural environment may contain inhibitory substances, (iii) the growth rate of the organism on the low ambient concentrations may be slower than the rate of predation, (iv) the added bacterium may use

Table 3. Pesticides and structurally related substances mentioned in the text.

Common or trade name	Class[a]	Chemical name
Alachlor	H	2-Chloro-N-(2,6-diethylphenyl)-N-(methoxymethyl) acetamide
Aldicarb	I,A,N	2-Methyl-2-(methylthio) propanal O-[(methylamino)carbonyl] oxime
Aldrin	I	1,2,3,4,10,10-Hexachloro-1,α4,α4aβ,5α,8α,8aβ-hexahydro-1,4:5,8-dimethanonaphthalene
Allyl alcohol	H	2-Propen-1-ol
Ametryne	H	N-Ethyl-N'-(1-methylethyl)-6-(methylthio)-1,3,5-triazine-2,4-diamine
Amitrole	H	1H-1,2,4-Triazol-3-amine
Asulam	H	Methyl[(4-aminophenyl)sulfonyl] carbamate
Atrazine	H	6-Chloro-N-ethyl-N'-(1-methylethyl)-1,3,5-triazine-2,4-diamine
Azinphos-methyl	I,A	O,O-dimethyl S-[(4-oxo-1,2,3-benzotriazin-3(4H)-yl)-methyl] phosphorodithioate
Barban	H	4-Chloro-2-butynyl 3-chlorophenylcarbamate
Baygon	I,M	2-(1-Methylethoxy)phenyl methylcarbamate
Benomyl	F	Methyl 1-[(butylamino)carbonyl]-1H-benzimidazol-2-ylcarbamate
Benthiocarb	H	S-[4-Chlorophenyl)methyl] diethylcarbamothioate
Bis(tributyltin) oxide	F	Hexabutyldistannoxane
Bromacil	H	5-Bromo-6-methyl-3-(1-methylpropyl)-2,4(1H,3H)-pyrimidinedione
Bromoxynil	H	3,5-Dibromo-4-hydroxybenzonitrile
Captafol	F	3a,4,7,7a-Tetrahydro-2-[(1,1,2,2-tetrachloroethyl)thio]-1H-isoindole-1,3(2H)-dione
Carbaryl	I	1-Naphthalenyl methylcarbamate
Carbendazim	F	Methyl 1H-benzimidazol-2-ylcarbamate
Carbofuran	I,A,N	2,3-Dihydro-2,2-dimethyl-7-benzofuranyl methylcarbamate
Chlorobromuron	H	N'-(4-bromo-3-chlorophenyl)-N-methoxy-N-methylurea
Chlordane	I	1,2,4,5,6,7,8,8-Octachloro-2,3,3a,4,7,7a-hexahydro-4,7-methano-1H-indene
Chlordimeform	A	N'-(4-Chloro-2-methylphenyl)-N,N-dimethyl methanimidamide
Chlorfenvinphos	I	2-Chloro-1-(2,4-dichlorophenyl) ethenyl diethyl phosphate
Chloridazon	H	5-Amino-4-chloro-2-phenyl-3(2H)-pyridazinone
Chlorpyrifos	I	O,O-Diethyl-O-(3,5,6-trichloro-2-pyridyl) phosphorothioate
Chlorpyrifos-methyl	I	O,O-Dimethyl-O-(3,5,6-trichloro-2-pyridyl) phosphorothioate
Chlortoluron	H	N'-(3-Chloro-4-methylphenyl)-N,N-dimethylurea

Table 3. (*Continued*)

Common or trade name	Class[a]	Chemical name
CIPC	H	1-Methylethyl 3-chlorophenylcarbamate
Coumaphos	I,N	*O*-(3-Chloro-4-methyl-2-oxo-2*H*-benzopyran-7-yl) *O,O*-diethyl phosphorothioate
Crag	H	2-(2,4-Dichlorophenoxy) ethyl hydrogen sulfate
Cycloate	H	*S*-Ethyl cyclohexylethyl carbamothioate
Cypermethrin	I	(*RS*)-Cyano(3-phenoxyphenyl)methyl (*IRS*)-*cis,trans*-3-(2,2-dichloroethenyl)-2,2-dimethylcycloprop-anecarboxylate
Dalapon	H	2,2-Dichloropropanoic acid
DDT	I	1,1′-(2,2,2-Trichloroethylidene) bis[4-chlorobenzene]
Decamethrin	I	(1*R*[1α(*S**),3α])-Cyano (3-phenoxyphenyl)methyl 3-(2,2-dibromoethenyl)-2,2-dimethylcyclopropanecarboxylate
Desmedipham	H	Ethyl [3-[[(phenylamino)carbonyl]oxy]phenyl]carbamate
Di-allate	H	*S*-(2,3-Dichloro-2-propenyl) bis(1-methylethyl)carbamothioate
Diazinon	I,A	*O,O*-Diethyl *O*-[6-methyl-2-(1-methylethyl)-4-pyrimidinyl phosphorothioate
Dicamba	H	3,6-Dichloro-2-methoxybenzoic acid
2,4-D	H	(2,4-Dichlorophenoxy) acetic acid
Dichlorprop	H	(±)-2-(2,4-Dichlorophenoxy) propanoic acid
Dichlorvos	I	2,2-Dichloroethenyl dimethyl phosphate
Dieldrin	I	3,4,5,6,9,9-Hexachloro-1aα,2β,2aα,3β,6β,6aα,7β-7aα-octahydro-2,7:3,6-dimethano naphth[2,3-b]oxirene
Diflubenzuron	I	*N*-[[(4-Chlorophenyl)amino]carbonyl]-2,6-difluorobenzamide
Diquat	H	6,7-Dihydrodipyrido[1,2-a:2′1′-C]pyrazinediium ion
Diuron	H	*N*′-(3,4-Dichlorophenyl)-*N,N*-dimethylurea
Endosulfan	I	6,7,8,9,10,10-Hexachloro-1,5,5a,6,9,9a-hexahydro-6,9-methano-2,4,3-benzodioxathiepin 3-oxide
Endothal	H	7-Oxabicyclo-(2.2.1)-heptane-2,3-dicarboxylic acid
EPTC	H	*S*-Ethyl-dipropylcarbamothioate
Ethion	I	*O,O,O′,O′*-Tetraethyl *S,S′*-methylene di(phosphorodithioate)
Fenac	H	2,3,6-Trichlorobenzeneacetic acid
Fenitrooxon	I	Dimethyl-*O*-(3-methyl-4-nitrophenyl) phosphate

Fenitrothion	I	*O,O*-Dimethyl *O*-(3-methyl-4-nitro-phenyl) phosphorothioate
Fensulfothion	I,N	*O,O*-Diethyl *O*-4-methylsulfinyl)phenyl phosphorothioate
Fenuron	H	*N,N*-Dimethyl-*N'*-phenylurea
Fenvalerate	I	(*RS*)-Cyano-3-phenoxyphenyl)methyl(*RS*)-4-chloro-α-(1-methylethyl) benzeneacetate
Fluvalinate	I	[α-Cyano-3-phenoxybenzyl-2-(2-chloro-4-(trifluoromethyl) anilino]-3-methylbutanoate
Fonofos	I	(±)-*O*-Ethyl *S*-phenyl ethylphosphonodithioate
Glyphosate	H	*N*-(Phosphonomethyl) glycine
Gamma HCH	I	1α,2α,3β,4α,5α,6β-Hexachlorocyclohexane
Heptachlor	I	1,4,5,6,7,8,8-Heptachloro-3a,4,7,7a-tetrahydro-4,7-methano-1*H*-indene
Hexachlorocyclohexane	I	1,2,3,4,5,6-Hexachlorocyclohexane
IPC	H	1-Methylethyl phenylcarbamate
Iprodione	F	3-(3,5-Dichlorophenyl)-*N*-(1-methylethyl)-2,4-dioxo-1-imidazolinecarboxamide
Isofenphos	I	1-Methylethyl 2-[[ethoxy[(1-methylethyl)amino]phosphinothioyl]oxy] benzoate
Kepone	I,F	1,1a,3,3a,4,5,5,5a,5b,6-Decachloro-octahydro-1,3,4-metheno-2*H*-cyclobuta[cd]pentalen-2-one
Lindane	I	1α,2α,3β,4α,5α,6β-Hexachlorocyclohexane
Linuron	H	*N'*-(3,4-Dichlorophenyl)-*N*-methoxy-*N*-methylurea
Malathion	A,M	Diethyl(dimethoxyphosphinothioyl) thiobutanedioate
Mancozeb	F	[[1,2-Ethanediylbis[carbamodithioato]](2⁻)]manganese with [[1,2-Ethanediylbis[car-bamodithioato]](2⁻)zinc
MCPA	H	(4-Chloro-2-methylphenoxy) acetic acid
MCPB	H	4-(4-Chloro-2-methylphenoxy) butanoic acid
Mecoprop	H	2-(4-Chloro-2-methylphenoxy) propanoic acid
Metamitron	H	4-Amino-3-methyl-6-phenyl-1,2,4-triazin-5(4*H*)-one
Methamidophos	I,A	*O,S*-Dimethyl phosphoramidothioate
Methomyl	I,N	Methyl *N*-[[(methylamino)carbonyl]oxy] ethanimidothioate
Methoxychlor	I	1,1'-(2,2,2-trichloroethylidene) bis[4-methoxybenzene]
Methyl parathion	I,M	*O,O*-Dimethyl O-4-nitrophenyl phosphorothioate
Metolachlor	H	2-Chloro-*N*-(2-ethyl-6-methylphenyl)-*N*-(2-methoxy-1-methylethyl) acetamide
Mirex	I	1,1a,2,2,3,3a,-4,5,5,5a,5b,6-dodecachlorooctahydro-1,3,4-metheno-1*H*-cyclobuta[cd] pentalene
Molinate	H	*S*-Ethyl hexahydro-1*H*-azepine-1-carbothioate
Monuron	H	*N'*-(4-Chlorophenyl)-*N,N*-dimethylurea
Oxamyl	I,N	Methyl 2-(dimethylamino)-*N*-[[(methyl-amino)carbonyl]oxy]-2-oxoethanimidothioate
Paraquat	H	1,1'-dimethyl-4,4'-bipyridinium ion

Table 3. (*Continued*)

Common or trade name	Class[a]	Chemical name
Parathion	I,M	*O,O*-Diethyl O-4-nitrophenyl phosphorothioate
Permethrin	I	(3-Phenoxyphenyl)methyl(*IRS*)-*cis*,*trans*-3-(2,2-dichloroethenyl)-2,2-dimethylcyclopropane carboxylate
Phenmedipham	H	3-[(Methoxycarbonyl)amino]phenyl(3-methylphenyl)carbamate
Phorate	I	*O,O*-Diethyl *S*-ethylthiomethyl phosphorodithioate
Picloram	H	4-Amino-3,5,6-trichloropyridine-2-carboxylic acid
Pirimiphos-ethyl	I	*O*-[2-(Diethylamino)-6-methyl-4-pyrimidinyl]*O,O*-diethyl phosphorothioate
Pirimiphos-methyl	I,A	*O*-[2-Diethylamino)-6-methyl-4-pyrimidinyl]*O,O*-dimethyl phosphorothioate
Profluralin	H	*N*-(Cyclopropylmethyl)-2,6-dinitro-N-propyl-4-(trifluoromethyl) benzenamine
Promecarb	I	3-Methyl-5-(1-methylethyl)phenyl methylcarbamate
2,4,5-TP	H	(±)-2-(2,4,5-Trichlorophenoxy) propanoic acid
Prometryn	H	*N,N'*-bis(1-methylethyl)-6-(methylthio)-1,3,5-triazine-2,4-diamine
Propachlor	H	2-Chloro-*N*-(1-methylethyl)-*N*-phenylacetamide
Propamocarb	F	Propyl 3-(dimethylamino)propylcarbamate
Propanil	H	*N*-(3,4-Dichlorophenyl)propanamide
Rotenone	I	[2,R-(2α,6aα,12aα)]-1,2,12,12a-Tetrahydro-8,9-dimethoxy-2-(1-methylethenyl)[1]benzopyrano[3,4-*b*]furo[2,3-*h*][1]benzopyran-6(6aH)- one
Swep	H	Methyl 3,4-dichlorophenylcarbamate
Terbufos	I	*S*-[[(1,1-Dimethylethyl)thio]methyl] *O,O*-diethyl phosphorodithioate
TMTD	F	Tetramethylthioperoxydicarbonic diamide
Triadimefon	F	1-(4-Chlorophenoxy)-3,3-dimethyl-1-(1*H*-1,2,4-triazol-1-yl)-2-butanone
Triadimenol	F	1-(4-Chlorophenoxy)-3,3-dimethyl-1-(1*H*-1,2,4-triazol-1-yl)butan-2-ol
Triallate	H	*S*-(2,3,3-Trichloro-2-propenyl)bis(1-methylethyl)carbamothioate
2,4,5-T	H	(2,4,5-Trichlorophenoxy)acetic acid
Trifluralin	H	2,6-Dinitro-*N,N*-dipropyl-4-(trifluoromethyl)benzenamine
Vinclozolin	F	3-(3,5-Dichlorophenyl)-5-ethenyl-5-methyl-2,4-oxazolidinedione
Warfarin	R	4-Hydroxy-3-(3-oxo-1-phenylbutyl)-2*H*-1-benzopyran-2-one

[a] A: acaricide; F: fungicide; H: herbicide; I: insecticide; M: miticide; N: nematicide; R: rodenticide.

the organic substrates in the environment rather than the pollutant, (v) the bacterium may fail to move through soil pores to sites of pollutant." They concluded that "the inoculum approach is feasible and worthwhile if means are found to overcome the ecological constraints on the inoculum strains", thus highlighting the lack of knowledge in microbial ecology which is vital to success in approaches to soil and water decontamination. Until funding agencies realize that this lack of knowledge is critical and start to provide adequate research funds for appropriate ecological studies, the discoveries that have arisen from the excellent funding provided for genetic studies in recent times might not be of use in environmental decontamination.

VI. Summary

This chapter provides a review concerning the microbial metabolism of pesticides and substances that are either major metabolites from pesticides or have structural similarity to certain pesticides, and covers the period 1981 to 1987. While reference has only been made to work published during this period, it should be realized that in some instances the results cited may confirm or expand upon earlier findings rather than being entirely novel. Therefore, the reader is referred to earlier reviews. The metabolism of pesticides in natural environments, water and wastewater, mixed microbial cultures, and pure cultures has been discussed. Attention has been drawn to the meager amount of information concerning the biodegradation of pesticides in anaerobic and marine environments. Issues such as the importance of cometabolism of pesticides in natural environments and a clear understanding of enhanced degradation of pesticides in soil still remain unresolved. Separate sections have been devoted to methodology in biodegradation studies, bound residues and removal of pesticides from soil and water. While pure culture studies have an important place in investigations into microbial metabolism of pesticides, increasing emphasis has been placed on the use of microbial consortia, either natural or artificial and microcosms to provide an understanding of pesticide biodegradation in natural environments. Another dimension in bound residue formation, one of physical entrapment in humic materials has been described. Various questions regarding the bioavailability of bound residues and whether they pose an environmental problem have not been answered fully. The microbiological removal of pesticides from soil and water by selected or genetically-engineered strains is discussed. It has been emphasized that the future success of such methods for the decontamination of soil and water depends very heavily on an improved knowledge of microbial ecology.

VII. References

Adhya TK, Barik S, Sethunathan N (1981a) Fate of fenitrothion, methyl parathion, and parathion in anoxic sulfur-containing soil systems. Pestic Biochem Physiol 16:14–20.

Adhya TK, Barik S, Sethunathan N (1981b) Hydrolysis of selected organophosphorus insecticides by two bacteria isolated from flooded soil. J Appl Bacteriol 50:167–172.

Adhya TK, Barik S, Sethunathan N (1981c) Stability of commercial formulation of fenitrothion, methyl parathion, and parathion in anaerobic soils. J Agric Food Chem 29:90–93.

Alexander M (1981) Biodegradation of chemicals of environmental concern. Science 211:132–138.

Alexander M (1985) Biodegradation of organic chemicals. Environ Sci Technol 19:106–111.

Allard A-S, Remberger M, Neilson AH (1987) Bacterial O-methylation of halogen-substituted phenols. Appl Environ Microbiol 53:839–845.

Amy PS, Schulke JW, Frazier LM, Seidler RJ (1985) Characterization of aquatic bacteria and cloning of genes specifying partial degradation of 2,4-dichlorophenoxyacetic acid. Appl Environ Microbiol 49:1237–1245.

Anderson JPE (1984) Herbicide degradation in soil: influence of microbial biomass. Soil Biol Biochem 16:483–489.

Anson JG, Mackinnon G (1984) Novel Pseudomonas plasmid involved in aniline degradation. Appl Environ Microbiol 48:868–869.

Antai SP, Crawford DL (1983) Degradation of phenol by *Streptomyces setonii*. Can J Microbiol 29:142–143.

Apajalahti JHA, Salkinoja-Salonen MS (1984) Absorption of pentachlorophenol (PCP) by bark chips and its role in microbial PCP degradation. Microbiol Ecol 10:359–367.

Aslanzadeh J, Hedrick HG (1985) Search for Mirex-degrading soil microorganisms. Soil Sci 139:369–374.

Attaway HH, Camper ND, Paynter MJB (1982) Anaerobic microbial degradation of diuron by pond sediment. Pestic Biochem Physiol 17:96–101.

Baarschers WH, Bharath AI, Elvish J, Davies M (1982) The biodegradation of methoxychlor by *Klebsiella pneumoniae*. Can J Microbiol 28:176–179.

Baarschers WH, Heitland HS (1986) Biodegradation of fenitrothion and fenitrooxon by the fungus *Trichoderma viride*. J Agric Food Chem 34:707–709.

Balthazor TM, Hallas LE (1986) Glyphosate-degrading microorganisms from industrial activated sludge. Appl Environ Microbiol 51:432–434.

Barik S, Munnecke DM (1982) Enzymatic hydrolysis of concentrated diazinon in soil. Bull Environ Contam Toxicol 29:235–239.

Barik S (1984) Metabolism of insecticides by microorganisms. In: Lal R (ed) Insecticide Microbiology. Springer-Verlag, Berlin pp. 87–128.

Bartholomew GW, Pfaender FK (1983) Influence of spatial and temporal variations on organic pollutant biodegradation rates in an estuarine environment. Appl Environ Microbiol 45:103–109.

Barug D (1981) Microbial degradation of bis(tributyltin)oxide. Chemosphere 10:1145–1154.

Bauer JE, Capone DG (1985) Degradation and mineralization of the polycyclic aromatic hydrocarbons anthracene and naphthalene in intertidal marine sediments. Appl Environ Microbiol 50:81–90.

Beeman RW, Matsumura F (1981) Metabolism of *cis*- and *trans*-chlordane by a soil micro-organism. J Agric Food Chem 29:84–88.

Behki RM, Khan SU (1986) Degradation of atrazine by *Pseudomonas*: N-Dealkylation and dehalogenation of atrazine and its metabolites. J Agric Food Chem 34:746–749.

Beltrame P, Beltrame PL, Carniti P (1984) Inhibiting action of chloro- and nitro-phenols on biodegradation of phenol: a structure-toxicity relationship. Chemosphere 13:3–9.

Benezet HJ, Knowles CO (1981) Degradation of chlordimeform by algae. Chemosphere 10:909–917.

Bengtsson G, Piwoni MD, Lundberg A (1986) Ultrafiltration cell for sorption and biodegradation experiments. Water Res 20:935–937.

Berry DF, Boyd SA (1984) Oxidative coupling of phenols and anilines by peroxidase: structure-activity relationships. Soil Sci Soc Am J 48:565–569.

Bollag J-M, Loll MJ (1983) Incorporation of xenobiotics into soil humus. Experientia 39:1221–1231.

Bollag, J-M, McGahen LL, Minard RD, Liu S-Y (1986) Bioconversion of alachlor in an anaerobic stream sediment. Chemosphere 15:153–162.

Bouwer EJ, McCarty PL (1982) Removal of trace chlorinated organic compounds by activated carbon and fixed-film bacteria. Environ Sci Technol 16:836–843.

Bouwer EJ, McCarty PL (1983) Transformations of halogenated organic compounds under denitrification conditions. Appl Environ Microbiol 45:1295–1299.

Boyd SA (1982) Adsorption of substituted phenols by soil. Soil Sci 134:337–343.

Boyd SA, Shelton DR, Berry D, Tiedje JM (1983) Anaerobic biodegradation of phenolic compounds in digested sludge. Appl Environ Microbiol 46:50–54.

Boyd SA, Shelton DR (1984) Anaerobic biodegradation of chlorophenols in fresh and acclimated sludge. Appl Environ Microbiol 47:272–277.

Brown EJ, Pignatello JJ, Martinson MM, Crawford RL (1986) Pentachlorophenol degradation: a pure bacterial culture and an epilithic microbial consortium. Appl Environ Microbiol 52:92–97.

Bull AT (1980) Biodegradation: some attitudes and strategies of microorganisms and microbiologists. In: Ellwood DC, Latham MJ, Slater JH, Lynch JM, Hedger JN (eds) Contemporary microbial ecology. Academic Press London, pp 107–136.

Bumpus JA, Aust SD (1987) Biodegradation of DDT [1,1,1-trichloro-2,2-bis(4-chloro-phenyl)ethane] by the white rot fungus *Phanerochaete chrysosporium*. Appl Environ Microbiol 53:2001–2008.

Bunce NJ, Merrick RL, Corke CT (1983) Reductive transformations of nitrate with 3,4-dichloroaniline and related compounds by *Escherichia coli*. J Agric Food Chem 31:1071–1075.

Capriel P, Haisch A, Khan SU (1986) Supercritical methanol: an efficacious technique for the extraction of bound pesticide residues from soil and plant samples. J Agric Food Chem 34:70–73.

Carr RJG, Bilton RF, Atkinson T (1985) Mechanism of biodegradation of paraquat by *Lipomyces starkeyi*. Appl Environ Microbiol 49:1290–1294.

Cerniglia CE, Lambert KJ, Miller DW, Freeman JP (1984) Transformation of 1- and 2-methylnaphthalene by *Cunninghamella elegans*. Appl Environ Microbiol 47:111–118.

Chapman RA, Tu CM, Harris CR, Cole C (1981) Persistence of five pyrethroid insecticides in sterile and natural, mineral and organic soil. Bull Environ Contam Toxicol 26:513–519.

Chapman RA, Tu CM, Harris CR, Dubois D (1982a) Biochemical and chemical transformations of terbufos, terbufos sulfoxide and terbufos sulfone in natural and sterile, mineral and organic soil. J Econ Entomol 75:955–960.

Chapman RA, Tu CM, Harris CR, Harris C (1982b) Biochemical and chemical transformations of phorate, phorate sulfoxide, and phorate sulfone in natural and sterile mineral and organic soil. J Econ Entomol 75:112–117.

Chapman RA, Tu CM, Harris CR, Harris C (1985) Persistence of diflubenzuron and Bay SIR8514 in natural and sterile sandy loam and organic soils. J Environ Sci Hlth Part B 20:489–497.

Chapman RA, Harris CR, Harris C (1986a) Observations on the effect of soil type, treatment in intensity, insecticide formulation, temperature and moisture on the adaptation and subsequent activity of biological agents associated with carbofuran degradation in soil. J Environ Sci Hlth Part B 21:125–141.

Chapman RA, Harris CR, Harris C (1986b) The effect of formulation and moisture level on the persistence of carbofuran in a soil containing biological systems adapted to its degradation. J Environ Sci Hlth Part B 21:57–66.

Chapman RA, Harris CR, Moy P, Henning K (1986c) Biodegradation of pesticides in soil: rapid degradation of isofenphos in a clay loam after a previous treatment. J Environ Sci Hlth Part B 21:269–276.

Chatterjee DK, Kilbane JJ, Chakrabarty AM (1982) Biodegradation of 2,4,5-trichlorophenoxyacetic acid in soil by a pure culture of *Pseudomonas cepacia*. Appl Environ Microbiol 44:514–516.

Cheng HH, Haider K, Harper SS (1983) Catechol and chlorocatechols in soil: degradation and extractability. Soil Biol Biochem 15:311–317.

Chesney RH, Sollitti P, Rubin HE (1985) Incorporation of phenol carbon at trace concentrations by phenol-mineralizing microorganisms in fresh water. Appl Environ Microbiol 49:15–18.

Chowdhury A, Vockel D, Moza PN, Klein W, Korte F (1981) Balance of conversion and degradation of 2,6-dichlorobenzonitrile-14C in humus soil. Chemosphere 10:1101–1108.

Cook AM, Hutter R (1981a) Triazines as nitrogen sources for bacteria. J Agric Food Chem 29:1135–1143.

Cook AM, Hutter R (1981b) Degradation of s-triazines: A critical view of biodegradation. In: Leisinger T, Hutter R, Cook AM, Nuesch J (eds) Microbial degradation of xenobiotics and recalcitrant compounds. Academic Press London, pp 237–249.

Cook AM, Hutter R (1982) Ametryne and prometryne as sulfur sources for bacteria. Appl Environ Microbiol 43:781–786.

Cook AM, Grossenbacher H, Hutter R (1983) Isolation and cultivation of microbes with biodegradative potential. Experientia 39:1191–1198.

Cook AM, Hutter R (1984) Deethylsimazine: bacterial dechlorination, deamination and complete degradation. J Agric Food Chem 32:581–585.

Corbett MD, Corbett BR (1981) Metabolism of 4-chloronitrobenzene by the yeast *Rhodosporidium* sp. Appl Environ Microbiol 41:942–949.

Coveney MF, Wetzel RG (1984) Improved double-vial radiorespirometric technique for mineralization of 14C-labeled substrates. Appl Environ Microbiol 47:1154–1157.

Cullen WR, McBride BC, Pickett AW, Reglinski J (1984) The wood preservative chromated copper arsenate is a substrate for trimethylarsine biosynthesis. Appl Environ Microbiol 47:443–444.

Dagley S (1983) Biodegradation and biotransformation of pesticides in the earth's carbon cycle. Residue Reviews 85:127–137.

Dalton H, Stirling DI (1982) Cometabolism. Philos Trans R Soc London B 297:481–496.

Davis PJ, Rizzo JD (1982) Microbial transformations of warfarin: stereoselective reduction by *Nocardia corallina* and *Arthrobacter* species. Appl Environ Microbiol 43:884–890.

Dean-Ross D (1987) Biodegradation of toxic wastes in soil. Am Soc Microbiol News 53:490–492.

Deas AHB, Clark T, Carter GA (1984a) The enantiomeric composition of triadimenol produced during metabolism of triadimefon by fungi. Part I Influence of dose and time of incubation. Pestic Sci 15:63–70.

Deas AHB, Clark T, Carter GA (1984b) The enantiomeric composition of triadimenol produced during metabolism of triadimefon by fungi. Part II Differences between fungal species. Pestic Sci 15:71–77.

DeBont JAM, Vorage MJAW, Hartmans S, van den Tweel WJJ (1986) Microbial degradation of 1,3-dichlorobenzene. Appl Environ Microbiol 52:677–680.

Deeley GM, Skierkowski P, Robertson JM (1985) Biodegradation of [14C]phenol in secondary sewage and landfill leachate measured by double-vial radio respirometry. Appl Environ Microbiol 49:867–869.

DeLaat J, Bouanga F, Dore M, Mallevialle J (1985) Influence of bacterial growth in granular activated carbon filters on the removal of biodegradable and of non-biodegradable organic compounds. Water Res 19:1565–1578.

DeLaune RD, Salinas LM (1985) Fate of 2,4-D entering a freshwater aquatic environment. Bull Environ Contam Toxicol 35:564–568.

Doneche B, Seguin G, Ribereau-Gayon P (1983) Mancozeb effect on soil microorganisms and its degradation in soils. Soil Sci 135:361–366.

Doyle RC, Kaufman DD, Burt GW, Douglass L (1981) Degradation of *cis*-permethrin in soil amended with sewage sludge or dairy manure. J Agric Food Chem 29:412–414.

Duah-Yentumi S, Kuwatsuka S (1982) Microbial degradation or benthiocarb, MCPA and 2,4-D herbicides in perfused soils amended with organic matter and chemical fertilizers. Soil Sci Plant Nutr 28:19–26.

Dwyer DF, Krumme ML, Boyd SA, Tiedje JM (1986) Kinetics of phenol biodegradation by an immobilized methanogenic consortium. Appl Environ Microbiol 52:345–351.

Eberbach PL, Douglas LA (1983) Persistence of glyphosate in a sandy loam. Soil Biol Biochem 15:485–487.

Edgehill RU, Finn RK (1983) Microbial treatment of soil to remove pentachlorophenol. Appl Environ Microbiol 45:1122–1125.

Ehrlich GG, Goerlitz DF, Godsy EM (1982) Degradation of phenolic contaminants in ground water by anaerobic bacteria. Ground Water 20:703–710.

Engelhardt G, Wallnofer PR, Rast HG (1981) Bacterial degradation of veratrylglycerol-β-arylethers as model compounds for soil-bound pesticide residues. In: Leisinger T, Hutter R, Cook AM, Nuesch J. (eds) Microbial degradation of xenobiotics and recalcitrant compounds. Academic Press, London, pp. 293–296.

Engelhardt G, Ziegler W, Wallnofer PR, Jarczyk HJ, Oehlmann L (1982) Degradation of the triazinone herbicide metamitron by *Arthrobacter* sp. DSM 20389. J Agric Food Chem 30:278–282.

Engelhardt G, Wallnofer PR (1983) Microbial transformation of benzazimide, a microbial degradation product of the insecticide azinphosmethyl. Chemosphere 12:955–960.

Fannin TE, Marcus MD, Anderson DA, Bergman HL (1981) Use of a fractional factorial design to evaluate interactions of environmental factors affecting biodegradation rates. Appl Environ Microbiol 42:936–943.

Felsot A, Maddox JV, Bruce W (1981) Enhanced microbial degradation of carbofuran in soils with histories of furadan use. Bull Environ Contam Toxicol 26:781–788.

Finn RK (1983) Use of specialized microbial strains in the treatment of industrial waste and in soil decontamination. Experientia 39:1231–1236.

Focht DD, Shelton D (1987) Growth kinetics of *Pseudomonas alcaligenes* C-O relative to inoculation and 3-chlorobenzoate metabolism in soil. Appl Environ Microbiol 53:1846–1849.

Fogel S, Lancione RL, Sewall AE (1982) Enhanced biodegradation of methoxychlor in soil under sequential environmental conditions. Appl Environ Microbiol 44:113–120.

Forrest M, Lord KA, Walker N (1981) The influence of soil treatments on the bacterial degradation of diazinon and other organophosphorus insecticides. Environ Pollut Ser A 24:93–104.

Fournier JC (1980) Enumeration of the soil microorganisms able to degrade 2,4-D by metabolism or co-metabolism. Chemosphere 9:169–174.

Fournier JC, Codaccioni P, Soulas G (1981) Soil adaptation to 2,4-D degradation in relation to the application rates and the metabolic behaviour of the degrading microflora. Chemosphere 10:977–984.

Fuhremann TW, Lichtenstein EP (1980) A comparative study of the persistence, movement, and metabolism of six carbon-14 insecticides in soils and plants. J Agric Food Chem 28:446–452.

Gerstl Z, Helling CS (1985) Fate of bound methyl parathion residues in soils as affected by agronomic practices. Soil Biol Biochem 17:667–673.

Getzin LW (1981) Degradation of chlorpyrifos in soil: influence of autoclaving, soil moisture and temperature. J Econ Entomol 74:158–162.

Ghisalba O (1983) Chemical wastes and their biodegradation—An Overview. Experientia 39:1247–1257.

Ghosal R, You I-S, Chatterjee DK, Chakrabarty AM (1985) Microbial degradation of halogenated compounds. Science (Washington, D.C.) 228:135–142.

Giardina MC, Giardi MT, Filacchioni G (1982) Atrazine metabolism by *Nocardia*: elucidation of initial pathway and synthesis of potential metabolites. Agric Biol Chem 46:1439–1445.

Gibson SA, Suflita JM (1986) Extrapolation of biodegradation results to groundwater aquifers: reductive dehalogenation of aromatic compounds. Appl Environ Microbiol 52:681–688.

Godsy EM, Goerlitz DF, Ehrlich GG (1986) Effects of pentachlorophenol on methanogenic fermentation of phenol. Bull Environ Contam Toxicol 36:271–277.

Goldstein RM, Mallory LM, Alexander M (1985) Reasons for possible failure of inoculation to enhance biodegradation. Appl Environ Microbiol 50:977–983.

Golovleva LA, Skryabin GK (1981) Microbial degradation of DDT. In: Leisinger T, Hutter R, Cook AM, Nuesch J (eds) Microbial degradation of xenobiotics and recalcitrant compounds. Academic Press, London, pp. 287–291.

Grossenbacher H, Horn C, Cook AM, Hutter R (1984) 2-Chloro-4-amino-1,3,5-triazine-6(5H)-one: A new intermediate in the biodegradation of chlorinated *s*-triazines. Appl Environ Microbiol 48:451–453.

Guenzi WD, Beard WE (1981) Degradation of *p*-chlorophenyl methyl sulfide, -sulfoxide, and -sulfone in soil. Soil Sci 131:135–139.

Guthrie MA, Kirsch EJ, Wukasch RF, Grady Jr CPL (1984) Pentachlorophenol biodegradation-II anaerobic. Water Res 18:451–461.

Haas D (1983) Genetic aspects of biodegradation by pseudomonads. Experientia 39: 1199–1213.

Hallas LE, Alexander M (1983) Microbial transformation of nitroaromatic compounds in sewage effluent. Appl Environ Microbiol 45:1234–1241.

Hannah SA, Austern BM, Eralp AE, Wise RH (1986) Comparative removal of toxic pollutants by six waste water treatment processes. J Water Pollut Control Fed 58:27–34.

Hardman DJ, Gowland PC, Slater JH (1986) Large plasmids from soil bacteria enriched on halogenated alkanoic acids. Appl Environ Microbiol 51:44–51.

Harris CR, Chapman RA, Harris C, Tu CM (1984) Biodegradation of pesticides in soil: rapid induction of carbamate-degrading factors after carbofuran treatment. J Environ Sci Hlth Part B 19:1–11.

Harvey Jr J (1983) A simple method of evaluating soil breakdown of 14C-pesticides under field conditions. Residue Reviews 85:149–158.

Hill NP, McIntyre AE, Perry R, Lester JN (1986) Behaviour of chlorophenoxy herbicides during the activated sludge treatment of municipal waste water. Water Res. 20:45–52.

Hoff T, Liu S-Y, Bollag J-M (1985) Transformation of halogen-, alkyl-, and alkoxy-substituted anilines by a laccase of *Trametes versicolor*. Appl Environ Microbiol 49:1040–1045.

Hoover DG, Borgonovi GE, Jones SH, Alexander M (1986) Anomalies in mineralization of low concentrations of organic compounds in lake water and sewage. Appl Environ Microbiol 51:226–232.

Horowitz A, Shelton DR, Cornell CR, Tiedje JM (1982) Anaerobic degradation of aromatic compounds in sediments and digested sludge. Dev Ind Microbiol 23:435–444.

Horowitz A, Suflita JM, Tiedje JM (1983) Reductive dehalogenations of halobenzoates by anaerobic lake sediment microorganisms. Appl Environ Microbiol 45:1459–1465.

Houx NWH, Dekker A (1987) A test system for the determining of the fate of pesticides in surface water. Int J Environ Anal Chem 29:37–59.

Huckins JN, Petty JD, England DC (1986) Distribution and impact of trifluralin, atrazine, and fonofos residues in microcosms simulating a northern prairie wetland. Chemosphere 15:563–588.

Hutchins SR, Tomson MB, Wilson JT, Ward CH (1984a) Microbial removal of waste water organic compounds as a function of input concentration in soil columns. Appl Environ Microbiol 48:1039–1045.

Hutchins SR, Tomson MB, Wilson JT, Ward CH (1984b) Anaerobic inhibition of trace organic compound removal during rapid infiltration of waste water. Appl Environ Microbiol 48:1046–1048.

Hutzinger O, Veerkamp W (1981) Xenobiotic chemicals with pollution potential. In: Leisinger T, Hutter R, Cook AM, Nuesch J (eds) Microbial degradation of xenobiotics and recalcitrant compounds. Academic Press, London. pp. 3–45.

Hwang H-M, Hodson RE, Lee RF (1985) Photochemical and microbial degradation of 2,4,5-trichloroaniline in a freshwater lake. Appl Environ Microbiol 50:1177–1180.

Jacob GS, Schaefer J, Stejskal EO, McKay RA (1985) Solid-state NMR determination of glyphosate metabolism in a *Pseudomonas* sp. J Biol Chem 360:5899–5905.

Jacobson SN, Alexander M (1981) Enhancement of the microbial dehalogenation of a model chlorinated compound. Appl Environ Microbiol 42:1062–1066.

Janke D, Fritsche W (1985) Nature and significance of microbial cometabolism of xenobiotics. J Basic Microbiol 25:603–619.

Jessee JA, Benoit RE, Hendricks AC, Allen GC, Neal JL (1983) Anaerobic degradation of cyanuric acid, cysteine, and atrazine by a facultative anaerobic bacterium. Appl Environ Microbiol 45:97–102.

Johnson LM, Williams FD (1982) Effect of substrate concentration on the cometabolism of m-chlorobenzoate by *Pseudomonas fluorescens*. Bull Environ Contam Toxicol 29:447–454.

Johnson LM, Talbot HW Jr (1983) Detoxification of pesticides by microbial enzymes. Experientia 39:1236–1246.

Jones SH, Alexander M (1986) Kinetics of mineralization of phenols in lake water. Appl Environ Microbiol 51:891–897.

Josephson J (1982) Supercritical fluids. Environ Sci Technol 16:548A–551A.

Josephson J (1983) Pesticides of the future. Environ Sci Technol 17:464A–468A.

Josephson J (1984a) Hazardous waste research. Environ Sci Technol 18:222A–223A.

Josephson J (1984b) Fixed film filtration. Environ Sci Technol 18:310A–313A.

Kaminski U, Janke D, Prauser H, Fritsche W (1983) Degradation of aniline and monochloroanilines by *Rhodococcus* sp. AN117 and a pseudomonad: a comparative study. Z Allg Mikrobiol 23:235–246.

Karns JS, Duttagupta S, Chakrabarty AM (1983a) Regulation of 2,4,5-trichlorophenoxyacetic acid and chlorophenol metabolism in *Pseudomonas cepacia* AC1100. Appl Environ Microbiol 46:1182–1186.

Karns JS, Kilbane JJ, Duttagupta S, Chakrabarty AM (1983b) Metabolism of halophenols by 2,4,5-trichlorophenoxyacetic acid-degrading *Pseudomonas cepacia*. Appl Environ Microbiol 46:1176–1181.

Kassim G, Stott DE, Martin JP, Haider K (1981) Stabilization and incorporation into biomass of phenolic and benzenoid carbons during biodegradation in soil. Soil Sci Soc

Am J 46:305–309.

Kearney PC, Karns JS, Muldoon MT, Ruth JM (1986) Coumaphos disposal by combined microbial and UV-ozonation reactions. J Agric Food Chem 34:702–706.

Kellogg ST, Chatterjee DK, Chakrabarty AM (1981) Plasmid-assisted molecular breeding: new technique for enhanced biodegradation of persistent toxic chemicals. Science 214:1133–1135.

Khan SU (1980) Pesticides in the soil environment. In: Wakeman RJ (ed) Fundamental aspects of pollution control and environmental sciences 5. Elsevier Scientific Publishing Co. Amsterdam pp. 163–197.

Khan SU (1982a) Studies on bound ^{14}C-prometryn residues in soil and plants. Chemosphere 11:771–795.

Khan SU (1982b) Bound pesticide residues in soil and plants. Residue Reviews 84:1–25.

Khan SU (1982c) Distribution and characteristics of bound residues of prometryn in an organic soil. J Agric Food Chem 30:175–179.

Khan SU, Hamilton HA (1980) Extractable and bound (non-extractable) residues of prometryn and its metabolites in an organic soil. J Agric Food Chem 28:126–132.

Khan SU, Ivarson KC (1981) Microbiological release of unextracted (bound) residues from an organic soil treated with prometryn. J Agric Food Chem 29:1301–1303.

Kiene RP, Capone DG (1986) Stimulation of methanogenesis by aldicarb and several other N-methyl carbamate pesticides. Appl Environ Microbiol 51:1247–1251.

Kilbane JJ (1986) Genetic aspects of toxic chemical degradation. Microbiol Ecol 12:135–145.

Kilbane JJ, Chatterjee DK, Karns JS, Kellogg ST, Chakrabarty AM (1982) Biodegradation of 2,4,5-trichlorophenoxyacetic acid by a pure culture of *Pseudomonas cepacia*. Appl Environ Microbiol 44:72–78.

Kilbane JJ, Chatterjee DK, Chakrabarty AM (1983) Detoxification of 2,4,5-trichlorophenoxyacetic acid from contaminated soil by *Pseudomonas cepacia*. Appl Environ Microbiol 45:1697–1700.

Kilpi S (1980) Degradation of some phenoxy acid herbicides by mixed cultures of bacteria isolated from soil treated with 2-(2-methyl-4-chloro) phenoxypropionic acid. Microbiol Ecol 6:261–270.

Kim CJ, Maier WJ (1986) Acclimation and biodegradation of chlorinated organic compounds in the presence of alternate substrates. J Water Pollut Control Fed 58:157–164.

Kirkpatrick D, Biggs SR, Conway B, Finn CM, Hawkins DR, Honda T, Ishida M, Powell GP (1981) Metabolism of N-(2,3-dichlorophenyl)-3,4,5,6-tetrachlorophthalamic acid (Techlofthalam) in paddy soil and rice. J Agric Food Chem 29:1149–1153.

Klecka GM, Maier WJ (1985) Kinetics of microbial growth on pentachlorophenol. Appl Environ Microbiol 49:46–53.

Knowles CO, Benezet HJ (1981) Microbial degradation of the carbamate pesticides desmedipham, phenmedipham, promecarb and propamocarb. Bull Environ Contam Toxicol 27:529–533.

Kobayashi H, Rittmann BE (1982) Microbial removal of hazardous organic compounds. Environ Sci Technol 16:170A–183A.

Koeppe MK, Lichtenstein EP (1984) Effects of organic fertilizers on the fate of ^{14}C-carbofuran in an agro-microcosm under soil run-off conditions. J Econ Entomol 77: 1116–1122.

Kool HJ (1984) Influence of microbial biomass on the biodegradability of organic compounds. Chemosphere 13:751–761.

Krause A, Hancock WG, Minard RD, Freyer AJ, Honeycutt RC, Le Baron HM, Paulson DL, Liu S-Y, Bollag J-M (1985) Microbial transformation of the herbicide metolachlor by a soil actinomycete. J Agric Food Chem 33:548–589.

Kunc F, Rybarova J (1983) Mineralization of carbon atoms of ^{14}C-2,4-D side chain and degradation ability of bacteria in soil. Soil Biol Biochem 15:141–144.

Kurihara N, Ohisa N, Nakajima M, Kakutani T, Senda M (1981) Relation between microbial degradation and polarographic half-wave potential of polychlorocyclohexenes and BHC isomers. Agric Biol Chem 45:1229–1235.

Lal R (1982) Accumulation, metabolism and effects of organophosphorus insecticides on microorganisms. Adv Appl Microbiol 28:149–200.

Lal R, Saxena DM (1982) Accumulation, metabolism and effect of organochlorine insecticides on microorganisms. Microbiol Rev 46:95–127.

Lammerding AM, Bunce NJ, Merrick RL, Corke CT (1982) Structural effects on the microbial diazotization of anilines. J Agric Food Chem 30:644–647.

Laplanche A, Bouvet M, Venien F, Martin G, Chabrolles A (1981) Modelling parathion changes in the natural environment laboratory experiment. Water Res 15:599–607.

Lappin HM, Greaves MP, Slater JH (1985) Degradation of the herbicide mecoprop [2-(2-methyl-4-chlorphenoxy) propionic acid] by a synergistic microbial community. Appl Environ Microbiol 49:429–433.

Larkin MJ, Day MJ (1985) The effect of pH on the selection of carbaryl-degrading bacteria from garden soil. J Appl Bacteriol 58:175–185.

Larkin MJ, Day MJ (1986) The metabolism of carbaryl by three bacterial isolates, *Pseudomonas* spp. (NCIB 12042 & 12043) and *Rhodococcus* sp. (NCIB 12038) from garden soil. J Appl Bacteriol 60:233–242.

Laskowski DA, Swann RL, McCall PJ, Bidlack HD (1983) Soil degradation studies. Residue Reviews 85:139–147.

Leatham GF, Crawford RL, Kirk TK (1983) Degradation of phenolic compounds and ring cleavage of catechol by *Phanerochaete chrysosporium*. Appl Environ Microbiol 46:191–197.

Lee A (1984) EPTC (S-ethyl N, N dipropylthiocarbamate)-degrading microorganisms isolated from a soil previously exposed to EPTC. Soil Biol Biochem 16:529–531.

Leisinger T (1983) Microorganisms and xenobiotic compounds. Experientia 39:1183–1191.

Leoni V, Hollick CB, D'Alessandro Deluca E, Collison RJ, Merolli S (1981) The soil degradation of chlorpyrifos and the significance of its presence in the superficial water in Italy. Agrochimica 25:414–426.

Lewis DL, Holm HW (1981) Rates of transformation of methyl parathion and diethyl phthalate by Aufwuchs microorganisms. Appl Environ Microbiol 42:698–703.

Lewis DL, Kollig HP, Hall TL (1983) Predicting 2,4-dichlorophenoxyacetic acid ester transformation rates in periphyton-dominated ecosystems. Appl Environ Microbiol 46:146–151.

Lewis DL, Hodson RE, Freeman III LF (1984) Effects of microbial community interactions on transformation rates of xenobiotic chemicals. Appl Environ Microbiol 48:561–565.

Lewis DL, Hodson RE, Freeman III LF (1985) Multiphasic kinetics for transformation of methyl parathion by *Flavobacterium* species. Appl Environ Microbiol 50:553–557.

Lichtenstein EP (1980) "Bound" residues in soils and transfer of soil residues in crops. Residue Reviews 76:147–153.

Lichtenstein EP, Liang TT, Koeppe MK (1982) Effects of fertilizers, captafol, and atrazine on the fate and translocation of [^{14}C] fonofos and [^{14}C] parathion in soil-plant microcosms. J Agric Food Chem 30:871–878.

Lichtenstein EP, Liang TT, Koeppe M (1983) Effects of soil mixing and flooding on the fate of metabolism of ^{14}C-fonofos and ^{14}C parathion in open and closed agricultural microcosms. J Econ Entomol 76:233–238.

Lieberman MT, Alexander M (1983) Microbial and nonenzymatic steps in the decomposition of dichlorvos (2,2-dichlorovinyl O,O-dimethyl phosphate). J Agric Food Chem 31:265–267.

Lillis V, Dodgson KS, White GF, Payne WJ (1983) Initiation of activation of a preemergent herbicide by a novel alkylsulfatase of *Pseudomonas putida* FLA. Appl Environ Microbiol 46:988–994.

Liu D, Strachan WMJ, Thomson K, Kwasniewska K (1981a) Determination of the biodegradability of organic compounds. Environ Sci Technol 15:788–793.

Liu D, Thomson K, Strachan WMJ (1981b) Biodegradation of carbaryl in simulated aquatic environment. Bull Environ Contam Toxicol 27:412–417.

Liu S-Y, Minard RD, Bollag J-M (1981c) Coupling reactions of 2,4-Dichlorophenol with various anilines. J Agric Food Chem 29:253–257.

Lyons CD, Katz S, Bartha R (1984) Mechanisms and pathways of aniline elimination from aquatic environments. Appl Environ Microbiol 48:491–496.

McBride KE, Kenny JW, Stalker DM (1986) Metabolism of the herbicide bromoxynil by *Klebsiella pneumoniae* subsp. *ozaenae*. Appl Environ Microbiol 52:325–330.

McCall PJ, Vrona SA, Kelley SS (1981) Fate of uniformly carbon-14 ring labeled 2,4,5-trichlorophenoxyacetic acid and 2,4-dichlorophenoxyacetic acid. J Agric Food Chem 29:100–107.

Macholz RM, Kujawa M (1985) Recent state of lindane metabolism Part III. Residue Reviews 94:119–149.

McKinley VL, Federle TW, Vestal JR (1983) Improvements in and environmental applications of double-vial radiorespirometry for the study of microbial mineralization. Appl Environ Microbiol 45:255–259.

MacRae IC, Evans SK (1983) Factors influencing the adsorption of bacteria to magnetite in water and wastewater. Water Res 17:271–277.

MacRae IC, Evans SK (1984) Removal of bacteria from water by adsorption to magnetite. Water Res 18:1377–1380.

MacRae IC (1985) Removal of pesticides in water by microbial cells adsorbed to magnetite. Water Res 19:825–830.

MacRae IC (1986a) Removal of chlorinated hydrocarbons from water and wastewater by bacterial cells adsorbed to magnetite. Water Res. 20:1149–1152.

MacRae IC (1986b) Formation and degradation of soil bound ^{14}C fenitrothion residues in two agricultural soils. Soil Biol Biochem 18:221–225.

MacRae IC, Yamaya Y, Yoshida T (1984) Persistence of hexachlorocyclohexane isomers in soil suspensions. Soil Biol Biochem 16:285–286.

MacRae IC, Cameron AJ (1985) Bacterial reduction of fensulfothion and its hydrolysis product 4-methylsulfinyl phenol. Appl Environ Microbiol 49:236–237.

Madsen EL, Alexander M (1985) Effects of chemical speciation on the minralization of organic compounds by microorganisms. Appl Environ Microbiol 50:342–349.

Maier-Bode H, Hartel K (1981) Linuron and monolinuron. Residue Reviews 77:1–352.

Marks TS, Smith ARW, Quirk AV (1984) Degradation of 4-chlorobenzoic acid by *Arthrobacter* sp. Appl Environ Microbiol 48:1020–1025.

Masunaga S, Urushigawa Y, Yonezawa Y (1986) Biodegradation pathway of *o*-cresol by heterogeneous culture. Water Res 20:477–484.

Matsumura F (1982) Degradation of pesticides in the environment by microorganisms and sunlight. In: Matsumura F and Krishnamurti CR (eds) Biodegradation of pesticides. Plenum Press, New York, pp. 67–87.

Maule A, Plyte S, Quirk AV (1987) Dehalogenation of organochlorine insecticides by mixed anaerobic microbial populations. Pestic Biochem Physiol 27:229–236.

Merritt GC, Watts JE, McDougall K (1981) *In vitro* degradation of organophosphorus insecticides by *Pseudomonas aeruginosa* isolated from fleece-rot lesions of sheep. Aust Vet J 57:531.

Mikesell MD, Boyd SA (1985) Reductive dechlorination of the pesticides 2,4-D,2,4,5-T and pentachlorophenol in anaerobic sludges. J Environ Qual 14:337–340.

Mikesell MD, Boyd SA (1986) Complete reductive dechlorination and mineralization of pentachlorophenol by anaerobic microorganisms. Appl Environ Microbiol 52: 861–865.

Miles CJ, Delfino JJ (1985) Fate of aldicarb, aldicarb sulfoxide and aldicarb sulfone in Floridan groundwater. J Agric Food Chem 33:455–460.

Miles JRW, Tu CM, Harris CR (1981) A laboratory study of the persistence of carbofuran and its 3-hydroxy-and 3-keto-metabolites in sterile and natural mineral and organic soils. J Environ Sci Hlth Part B. 16:409–417.

Miles JRW, Moy P (1982) Degradation of the insecticide fensulfothion by a mixed culture of soil microorganisms. J Environ Sci Hlth Part B. 17:675–681.

Miles JRW, Harris CR, Tu CM (1984) Influence of moisture on the persistence of chlorpyrifos and chlorfenvinphos in sterile and natural mineral and organic soils. J Environ Sci Hlth Part B 19:237–243.

Milner CR, Goulder R (1986) The abundance, heterotrophic activity and taxonomy of bacteria in a stream subject to pollution by chlorophenols, nitrophenols and phenoxyalkanoic acids. Water Res 20:85–90.

Minard RD, Liu SY, Bollag J-M (1981) Oligomers and quinones from 2,4-dichlorophenol. J Agric Food Chem 29:250–253.

Molin G, Nilsson I (1985) Degradation of phenol by *Pseudomonas putida* ATCC 11172 in continuous culture at different ratios of biofilm surface to culture volume. Appl Environ Microbiol 50:946–950.

Moore JC, Hansen DJ, Garnas RL, Goodman LR (1985) A sand-granular carbon filtration treatment system for removing aqueous pesticide residues from a marine toxicology laboratory effluent. Water Res 19:1601–1604.

Moore JK, Braymer HD, Larson AD (1983) Isolation of a *Pseudomonas* sp. which utilizes the phosphonate herbicide glyphosate. Appl Environ Microbiol 46:316–320.

Moos IP, Kirsch EJ, Wukasch RF, Grady Jr CPL (1983) Pentachlorophenol biodegradation—I aerobic. Water Res 17:1575–1584.

Motosugi K, Soda K (1983) Microbial degradation of synthetic organochlorine compounds. Experientia 39:1214–1220.

Mount ME, Oehme FW (1981) Carbaryl: a literature review. Residue Reviews 80:1–64.

Mulbry WW, Karns JS, Kearney PC, Nelson JO, McDaniel CS, Wild JR (1986) Identification of a plasmid-borne parathion hydrolase gene from *Flavobacterium* sp. by southern hybridization with pod from *Pseudomonas diminuta*. Appl Environ Microbiol 51:926–930.

Mulla MS, Mian LS, Kawecki JA (1981) Distribution, transport, and fate of the insecticides malathion and parathion in the environment. Residue Reviews 81:1–159.

Munnecke DM (1981) The use of microbial enzymes for pesticide detoxification. In: Leisinger T, Hutter R, Cook AM, Nuesch J (eds) Microbial degradation of xenobiotics and recalcitrant compounds. Academic Press, London. pp. 251–269.

Murphy SE, Drotar AM, Fall R (1982) Biotransformation of the fungicide pentachloronitrobenzene by *Tetrahymena thermophila*. Chemosphere 11:33–39.

Neilson AH, Allard A-S, Hynning P-A, Remberger M, Landner L (1983) Bacterial methylation of chlorinated phenols and guaiacols: Formation of veratroles from guaiacols and high-molecular-weight chlorinated lignin. Appl Environ Microbiol 45:774–783.

Nelson LM (1982) Biologically-induced hydrolysis of parathion in soil: isolation of hydrolyzing bacteria. Soil Biol Biochem 14:219–222.

Nelson LM, Yaron B, Nye PH (1982) Biologically-induced hydrolysis of parathion in soil: kinetics and modelling. Soil Biol Biochem 14:223–227.

Nimmo WB, De Wilde PC, Verloop A (1984) The degradation of diflubenzuron and its chief metabolites in soils. Part I: hydrolytic cleavage of diflubenzuron. Pestic Sci 15:574–585.

Nimmo WB, Willems AGM, Joustra KD, Verloop A (1986) The degradation of diflubenzuron and its chief metabolites in soils. Part II Fate of 4-chlorophenylurea. Pestic Sci 17:403–411.

Norris LA (1981) The movement, persistence and fate of the phenoxy herbicides and TCDD in the forest. Residue Reviews 80:65–135.

Novick NJ, Alexander M (1985) Cometabolism of low concentrations of propachlor, alachlor, and cycloate in sewage and lake water. Appl Environ Microbiol 49:737–743.

Novick NJ, Mukherjee R, Alexander M (1986) Metabolism of alachlor and propachlor in suspensions of pretreated soils and in samples from ground water aquifers. J Agric Food Chem 34:721–725.

Ogram AV, Jessup RE, Ou LT, Rao PSC (1985) Effects of sorption on biological degradation rates of (2,4-dichlorophenoxy) acetic acid in soils. Appl Environ Microbiol 49:582–587.

Ohisa N, Yamaguchi M, Kurihara N (1980) Lindane degradation by cell-free extracts of

Clostridium rectum. Arch Microbiol 125:221–225.

Ohisa N, Kurihara N, Nakajima M (1982) ATP synthesis associated with the conversion of hexachlorocyclohexane related compounds. Arch Microbiol 131:330–333.

Omenn GS (1986) Genetic control of environmental pollutants: a conference review. Microbiol Ecol 12:129–134.

Ou L-T (1984) 2,4-D Degradation and 2,4-D degrading microorganisms in soils. Soil Sci 137:100–107.

Ou L-T (1985) Methyl parathion degradation and metabolism in soil: influence of high soil-water contents. Soil Biol Biochem 17:241–243.

Ou L-T, Rao PSC, Davidson JM (1983) Methyl parathion degradation in soil: influence of soil-water tension. Soil Biol Biochem 15:211–215.

Ou L, Edvarson KSV, Rao PSC (1985) Aerobic and anaerobic degradation of aldicarb in soils. J Agric Food Chem 33:72–78.

Padhy RN (1985) Cyanobacteria and pesticides. Residue Reviews 95:1–44.

Painter HA, King EF (1986) The need for applying stability tests in biodegradability assessments. Chemosphere 15:471–478.

Paris DF, Steen WC, Baughman GL, Barnett Jr JT (1981) Second-order model to predict microbial degradation of organic compounds in natural waters. Appl Environ Microbiol 41:603–609.

Paris DF, Wolfe NL, Steen WC (1982) Structure-activity relationships in microbial transformation of phenols. Appl Environ Microbiol 44:153–158.

Paris DF, Wolfe NL, Steen WC, Baughman GL (1983) Effect of phenol molecular structure on bacterial transformation rate constants in pond and river samples. Appl Environ Microbiol 45:1153–1155.

Paris DF, Wolfe NL, Steen WC (1984) Microbial transformation of esters of chlorinated carboxylic acids. Appl Environ Microbiol 47:7–11.

Paris DF, Wolfe NL (1987) Relationship between properties of a series of anilines and their transformation by bacteria. Appl Environ Microbiol 53:911–916.

Parker LW, Doxtader KG (1983) Kinetics of the microbial degradation of 2,4-D in soil: effects of temperature and moisture. J Environ Qual 12:553–558.

Pemberton JM (1981) Genetic engineering and biological detoxification of environmental pollutants. Residue Reviews 78:1–11.

Pemberton JM, Wynne EC (1984) Genetic engineering and biological detoxification/degradation of pesticides. In: Lal R (ed) Insecticide microbiology, Springer-Verlag, Berlin, pp 147–168.

Pettigrew CA, Paynter MJB, Camper ND (1985) Anaerobic microbial degradation of the herbicide propanil. Soil Biol Biochem 17:815–818.

Pfaender FK, Bartholomew GW (1982) Measurement of aquatic biodegradation rates by determining heterotrophic uptake of radiolabeled pollutants. Appl Environ Microbiol 44:159–164.

Pignatello JJ (1986) Ethylene dibromide mineralization in soils under aerobic conditions. Appl Environ Microbiol 51:588–592.

Pignatello JJ, Martinson MM, Steiert JG, Carlson RE, Crawford RL (1983) Biodegradation and photolysis of pentachlorophenol in artificial freshwater streams. Appl Environ Microbiol 46:1024–1031.

Pignatello JJ, Johnson LK, Martinson MM, Carlson RE, Crawford RL (1986) Response of the microflora in outdoor experimental streams to pentachlorophenol: environmental factors. Can J Microbiol 32:38–46.

Pillai P, Helling CS, Dragun J (1982) Soil-catalyzed oxidation of aniline. Chemosphere 11:299–317.

Pipke R, Schulz A, Amrhein N (1987) Uptake of glyphosate by an *Arthrobacter* sp. Appl Environ Microbiol 53:974–978.

Pore RS, Sorenson WG (1981) Kepone removal from aqueous solution by immobilized algae. J Environ Sci Hlth Part A 16:51–63.

Rajagopal BS, Chendrayan K, Reddy BR, Sethunathan N (1983) Persistence of carbaryl in flooded soils and its degradation by soil enrichment cultures. Plant Soil 73: 35–45.

Rajagopal BS, Brahmaprakash GP, Reddy BR, Singh UD, Sethunathan N (1984a) Effect and persistence of selected carbamate pesticides in soil. Residue Reviews 93: 1–199.

Rajagopal BS, Rao VR, Nagendrappa G, Sethunathan N (1984b) Metabolism of carbaryl and carbofuran by soil-enrichment and bacterial cultures. Can J Microbiol 30: 1458–1466.

Rajagopal BS, Sethunathan N (1984) Influence of nitrogen fertilizers on the persistence of carbaryl and carbofuran in flooded soils. Pestic Sci 15:591–599.

Rajagopal BS, Panda S, Sethunathan N (1986) Accelerated degradation of carbaryl and carbofuran in a flooded soil pretreated with hydrolysis products, 1-naphthol and carbofuran phenol. Bull Environ Contam Toxicol 36:827–832.

Ralph W (1986) A new generation of insecticides. Rural Res 133:4–7.

Reddy BR, Sethunathan N (1983) Mineralization of parathion in the rice rhizosphere. Appl Environ Microbiol 45:826–829.

Reddy BR, Sethunathan N (1985) Salinity and persistence of parathion in flooded soils. Soil Biol Biochem 17:235–239.

Reineke W, Knackmuss H-J (1984) Microbial metabolism of haloaromatics: isolation and properties of a chlorobenzene-degrading bacterium. Appl Environ Microbiol 47:395–402.

Reinert KH, Rodgers Jr JH, Leslie TJ, Hinman ML (1986) Static shake-flask biotransformation of endothall. Water Res 20:255–258.

Rosenberg A (1984) 2,3,6-Trichlorophenylacetic acid (fenac) degradation in aqueous and soil systems. Bull Environ Contam Toxicol 32:383–390.

Rubin HE, Subba-Rao RV, Alexander M (1982) Rates of mineralization of trace concentrations of aromatic compounds in lake water and sewage samples. Appl Environ Microbiol 43:1133–1138.

Rubin HE, Alexander M (1983) Effect of nutrients on the rates of mineralization of trace concentrations of phenol and *p*-nitrophenol. Environ Sci Technol 17:104–107.

Rubin HE, Schmidt S (1985) Growth of phenol-mineralizing microorganisms in fresh water. Appl Environ Microbiol 49:11–14.

Ruckdeschell G, Renner G (1986) Effects of pentachlorophenol and some of its known and possible metabolites on fungi. Appl Environ Microbiol 51:1370–1372.

Saber DL, Crawford RL (1985) Isolation and characterization of *Flavobacterium* strains that degrade pentachlorophenol. Appl Environ Microbiol 50:1512–1518.

Sahm H, Brunner M, Schoberth SM (1986) Anaerobic degradation of halogenated aromatic compounds. Microbiol Ecol 12:147–153.

Salkinoja-Salonen MS, Hakulinen R, Valo R, Apajalahti J (1983) Biodegradation of recalcitrant organochlorine compounds in fixed film reactors. Water Sci Technol 15:309–319.

Sariaslani FS, Rosazza JP (1985) Biotransformations of 1', 2'-dihydrorotenone by *Streptomyces griseus*. Appl Environ Microbiol 49:451–452.

Sattar MA (1981) Persistence of 4-chloro-*o*-cresol and 5-chloro-3-methylcatechol in soil. Chemosphere 10:1011–1017.

Saxena A, Bartha R (1983a) Microbial mineralization of humic acid-3,4-dichloroaniline complexes. Soil Biol Biochem 15:59–62.

Saxena A, Bartha R (1983b) Binding of 3,4-dichloroaniline by humic acid and soil: mechanism and exchangeability. Soil Sci 136:111–116.

Saxena A, Zhang R, Bollag J-M (1987) Microorganisms capable of metabolizing the herbicide metolachlor. Appl Environ Microbiol 53:390–396.

Schimmel S, Garnas RL, Patrick JM Jr, Moore JC (1983) Acute toxicity, bioconcentration, and persistence of AC 222,705, benthiocarb, chlorpyrifos, fenvalerate, methyl parathion and permethrin in the estuarine environment. J Agric Food Chem 31:104–113.

Schmidt E, Hellwig M, Knackmuss H-J (1983) Degradation of chlorophenols by a defined mixed microbial community. Appl Environ Microbiol 46:1038–1044.

Schmidt SK, Alexander M (1985) Effects of dissolved organic carbon and second substrates on the biodegradation of organic compounds at low concentrations. Appl Environ Microbiol 49:822–827.

Schmidt SK, Simkins S, Alexander M (1985) Models for the kinetics of biodegradation of organic compounds not supporting growth. Appl Environ Microbiol 50:323–331.

Schwien U, Schmidt E (1982) Improved degradation of monochlorophenols by a constructed strain. Appl Environ Microbiol 44:33–39.

Serdar CM, Gibson DT, Munnecke DM, Lancaster JH (1982) Plasmid involvement in parathion hydrolysis by *Pseudomonas diminuta*. Appl Environ Microbiol 44:246–249.

Sethunathan N, Adhya TK, Raghu K (1982) Microbial degradation of pesticides in tropical soils In: Matsumura F and Krishnamurti CR (eds) Biodegradation of pesticides, Plenum Press, New York, pp. 91–115.

Shaler TA, Klecka GM (1986) Effects of dissolved oxygen concentration on biodegradation of 2,4-dichlorophenoxyacetic acid. Appl Environ Microbiol 51:950–955.

Sheela S, Pai SB (1983) Metabolism of fensulfothion by a soil bacterium, *Pseudomonas alcaligenes* C_1. Appl Environ Microbiol 46:475–479.

Shelton DR, Tiedje JM (1984a) General method for determining anaerobic biodegradation potential. Appl Environ Microbiol 47:850–857.

Shelton DR, Tiedje JM (1984b) Isolation and partial characterization of bacteria in an anaerobic consortium that mineralizes 3-chlorobenzoic acid. Appl Environ Microbiol 48:840–848.

Shimp RJ, Pfaender FK (1985a) Influence of easily degradable naturally occurring carbon substrates on biodegradation of monosubstituted phenols by aquatic bacteria. Appl Environ Microbiol 49:394–401.

Shimp R, Pfaender FK (1985b) Influence of naturally occurring humic acids on biodegradation of monosubstituted phenols by aquatic bacteria. Appl Environ Microbiol 49:402–407.

Shimp RJ, Pfaender FK (1987) Effect of adaptation to phenol on biodegradation of monosubstituted phenols by aquatic microbial communities. Appl Environ Microbiol 53:1496–1499.

Shinabarger DL, Schmitt EK, Braymer HD, Larson AD (1984) Phosphonate utilization by the glyphosate-degrading *Pseudomonas* sp. strain PG2982. Appl Environ Microbiol 48:1049–1050.

Shirkot CK, Gupta KG (1985) Accelerated tetramethylthiuram disulfide (TMTD) degradation in soil by inoculation with TMTD-utilizing bacteria. Bull Environ Contam Toxicol 35:354–361.

Simkins S, Alexander M (1984) Models for mineralization kinetics with the variables of substrate concentration and population density. Appl Environ Microbiol 47:1299–1306.

Simkins S, Mukherjee R, Alexander M (1986) Two approaches to modeling kinetics of biodegradation by growing cells and application of a two-compartment model for mineralization kinetics in sewage. Appl Environ Microbiol 51:1153–1160.

Sims R, Sorensen D, Sims J, McLean J, Mahmood R, DuPont R, Jurinak J, Wagner K (1986) Contaminated surface soils in-place treatment techniques. Noyes Publications, Park Ridge, N.J.

Singh GJP (1981) Studies on the role of microorganisms in the metabolism of dieldrin in the epicuticular wax layer of blowflies *Calliphora erythrocephala*. Pestic Biochem Physiol 16:256–266.

Sjoblad RD, Bollag J-M (1981) Oxidative coupling of aromatic compounds by enzymes from soil microorganisms. In: Paul EA, Ladd JN (eds) Soil biochemistry. Marcel Dekker, Inc. New York pp. 113–152.

Slater JH, Bull AT (1982) Environmental microbiology biodegradation. Philos Trans R Soc London B 297:575–597.

Sleat R, Robinson JP (1984) The bacteriology of anaerobic degradation of aromatic compounds. J Appl Bacteriol 57:381–394.

Smith AE (1985a) Identification of 3,4-dichloroanisole and 2,4-dichlorophenol as soil degradation products of ring labelled [^{14}C]2,4-D. Bull Environ Contam Toxicol 34:150–157.

Smith AE (1985b) Identification of 4-chloro-2-methylphenol as a soil degradation product of ring-labelled [^{14}C] mecoprop. Bull Environ Contam Toxicol 34:656–660.

Smith AE, Hayden BJ (1981) Relative persistence of MCPA, MCPB and mecoprop in Saskatchewan soils and the identification of MCPA in MCPB-treated soils. Weed Res 21:179–183.

Smith S, Willis GH (1985) Disappearance of residual fenvalerate from sugar-cane trash as affected by trash placement in soil. Soil Sci 139:553–557.

Smolenski WJ, Suflita JM (1987) Biodegradation of cresol isomers in anoxic aquifers. Appl Environ Microbiol 53:710–716.

Somerville CC, Monti CA, Spain JC (1985) Modification of the ^{14}C most-probable-number method for use with nonpolar and volatile substrates. Appl Environ Microbiol 49:711–713.

Soulas G (1982) Mathematical model for microbial degradation of pesticides in the soil. Soil Biol Biochem 14:107–115.

Soulas G, Fournier JC (1981) Soil aggregate as a natural sampling unit for studying behaviour of microoganisms in the soil: application to pesticide degrading microorganisms. Chemosphere 10:431–440.

Soulas G, Codaccioni P, Fournier JC (1983) Effect of cross-treatment on the subsequent breakdown of 2,4-D, MCPA and 2,4,5-T in the soil. Behaviour of the degrading microbial populations. Chemosphere 12:1101–1106.

Soulas G, Chaussod R, Verguet A (1984) Chloroform fumigation technique as a means of determining the size of specialized soil microbial populations: application to pesticide-degrading microorganisms. Soil Biol Biochem 16:497–501.

Spain JC, Van Veld PA (1983) Adaptation of natural microbial communities to degradation of xenobiotic compounds: effects of concentration, exposure, time, inoculum, and chemical structure. Appl Environ Microbiol 45:428–435.

Spain JC, Van Veld PA, Monti CA, Pritchard PH, Cripe CR (1984) Comparison of p-nitrophenol biodegradation in field and laboratory test systems. Appl Environ Microbiol 48:944–950.

Spain JC, Nishino SF (1987) Degradation of 1,4-dichlorobenzene by a *Pseudomonas* sp. Appl Environ Microbiol 53:1010–1019.

Speece RE (1983) Anaerobic biotechnology for industrial waste water treatment. Environ Sci Technol 17:416A–427A.

Staiger LE, Quistad GB (1983) Degradation and movement of fluvalinate in soil. J Agric Food Chem 31:599–603.

Stanlake GJ, Finn RK (1982) Isolation and characterization of a pentachlorophenol-degrading bacterium. Appl Environ Microbiol 44:1421–1427.

Steiert JG, Pignatello JJ, Crawford RL (1987) Degradation of chlorinated phenols by a pentachlorophenol-degrading bacterium. Appl Environ Microbiol 53:907–910.

Stepp TD, Camper ND, Paynter MJB (1985) Anaerobic microbial degradation of selected 3,4-dihalogenated aromatic compounds. Pestic Biochem Physiol 23:256–260.

Stevenson FJ (1976) Organic matter reactions involving pesticides in soil. In: Kaufman DD, Still GG, Paulson GD, Bandal SK (eds) Bound and conjugated pesticide residues. American Chemical Society, Washington, D.C., pp 180–200.

Stott DE, Martin JP, Focht DD, Haider K (1983) Biodegradation, stabilization in humus, and incorporation into soil biomass of 2,4-D and chlorocatechol carbons. Soil Sci Soc Am J 47:66–70.

Stotzky G, Babich H (1986) Survival of, and genetic transfer by genetically engineered bacteria in natural environments. Adv Appl Microbiol 31:93–138.

Stralka KA, Camper ND (1981) Microbial degradation of profluralin. Soil Biol Biochem 13:33–38.

Stucki G, Alexander M (1987) Role of dissolution rate and solubility in biodegradation of aromatic compounds. Appl Environ Microbiol 53:292–297.

Subba-Rao RV, Alexander M (1985) Bacterial and fungal cometabolism of 1,1,1-trichloro-2, 2-bis(4-chlorphenyl)ethane(DDT) and its breakdown products. Appl Environ Microbiol 49:509–516.

Subba-Rao RV, Rubin HE, Alexander M (1982) Kinetics and extent of mineralization of organic chemicals at trace levels in freshwater and sewage. Appl Environ Microbiol

43:1139–1150.

Suflita JM, Bollag J-M (1981) Polymerization of phenolic compounds by a soil-enzyme complex. Soil Sci Soc Am J 45:297–302.

Suflita JM, Horowitz A, Shelton DR, Tiedje JM (1982) Dehalogenation: a novel pathway for the anaerobic biodegradation of haloaromatic compounds. Science 218: 1115–1117.

Suflita JM, Robinson JA, Tiedje JM (1983) Kinetics of microbial dehalogenation of haloaromatic substrates in methanogenic environments. Appl Environ Microbiol 45:1466–1473.

Suflita JM, Stout J, Tiedje JM (1984) Dechlorination of (2,4,5-trichlophenoxy) acetic acid by anaerobic microorganisms. J Agric Food Chem 32:218–221.

Suflita JM, Smolenski WJ, Robinson JA (1987) Alternative nonlinear model for estimating second-order rate coefficients for biodegradation. Appl Environ Microbiol 53:1064–1068.

Tam AC, Behki RM, Khan SU (1987) Isolation and characterization of an s-ethyl-N, N-dipropylthiocarbamate-degrading *Arthrobacter* strain and evidence for plasmid-associated s-ethyl-N, N-dipropylthiocarbamate degradation. Appl Environ Microbiol 53:1088–1093.

Thomas JM, Yordy JR, Amador JA, Alexander M (1986) Rates of dissolution and biodegradation of water-insoluble organic compounds. Appl Environ Microbiol 52:290–296.

Timms P, MacRae IC (1982) Conversion of fensulfothion by *Klebsiella pneumoniae* to fensulfothion sulfide and its accumulation. Aust J Biol Sci 35:661–667.

Timms P, MacRae IC (1983) Reduction of fensulfothion and accumulation of the product, fensulfothion sulfide, by selected microbes. Bull Environ Contam Toxicol 31:112–115.

Torstensson L, Wessen B (1984) Interactions between the fungicide benomyl and soil microorganisms. Soil Biol Biochem 16:445–452.

Trevors JT (1982) Effect of temperature on the degradation of pentachlorophenol by *Pseudomonas* species. Chemosphere 11:471–475.

van den Tweel WJJ, Kok JB, de Bont JAM (1987) Reductive dechlorination of 2,4-dichlorobenzoate to 4-chlorobenzoate and hydrolytic dehalogenation of 4-chloro-, 4-bromo-, and 4-iodobenzoate by *Alcaligenes denitrificans* NTB-1. Appl Environ Microbiol 53:810–815.

Van Der Kooij D, Visser A, Oranje JP (1982) Multiplication of fluorescent pseudomonads at low substrate concentrations in tap water. Antonie Leeuwenhoek 48:229–243.

Van Veld PA, Spain JC (1983) Degradation of selected xenobiotic compounds in three types of aquatic test systems. Chemosphere 12:1291–1305.

Vandenbergh PA, Olsen RH, Colaruotolo JF (1981) Isolation and genetic characterization of bacteria that degrade chloroaromatic compounds. Appl Environ Microbiol 42:737–739.

Veerkamp W, Pel R, Hutzinger O (1983) Transformation of chlorobenzoic acids by a *Pseudomonas* spec.: comparison of batch and chemostat cultures. Chemosphere 12: 1337–1343.

Vega D, Bastide J, Coste C (1985) Isolation from soil and growth characteristics of a CIPC-degrading strain of *Pseudomonas cepacia*. Soil Biol Biochem 17:541–545.

Venkateswarlu K, Sethunathan N (1984) Degradation of carbofuran by *Azospirillum lipoferum* and *Streptomyces* spp. isolated from flooded alluvial soil. Bull Environ Contam Toxicol 33:556–560.

Vankateswarlu K, Sethunathan N (1985) Enhanced degradation of carbofuran by *Pseudomonas cepacia* and *Nocardia* sp. in the presence of growth factors. Plant Soil 84:445–449.

Walker A, Brown PA, Entwistle AR (1986) Enhanced degradation of iprodione and vinclozolin in soil. Pestic Sci 17:183–193.

Wallnofer PR, Ziegler W, Engelhardt G (1981) Bacterial oxidation of 4-methylthio-substituted veratrylglycerol-β-aryl-ethers, model compounds for soil bound pesticide residues. Chemosphere 10:341–346.

Wang Y-S, Subba-Rao RV, Alexander M (1984) Effect of substrate concentration and organic and inorganic compounds on the occurrence and rate of mineralization and cometabolism. Appl Environ Microbiol 47:1195–1200.

Weinberger P, Greenhalgh R, Moody RP, Boulton B (1982a) Fate of fenitrothion in aquatic microcosms and the role of aquatic plants. Environ Sci Technol 16:470–473.

Weinberger P, Greenhalgh R, Sher D, Ouellette M (1982b) Persistence of formulated fenitrothion in distilled, estuarine, and lake water microcosms in dynamic and static systems. Bull Environ Contam Toxicol 28:484–489.

Weiss UM, Scheunert I, Klein W, Korte F (1982) Fate of pentachlorophenol-^{14}C in soil under controlled conditions. J Agric Food Chem 30:1191–1194.

Wiggins BA, Jones SH, Alexander M (1987) Explanations for the acclimation period preceding the mineralization of organic chemicals in aquatic environments. Appl Environ Microbiol 53:791–796.

Wood JM (1982) Chlorinated hydrocarbons: Oxidation in the biosphere. Environ Sci Technol 16:291A–297A.

Wright SJL, Maule A (1982) Transformation of the herbicides propanil and chlorpropham by micro-algae. Pestic Sci 13:253–256.

You I-S, Bartha R (1982a) Stimulation of 3,4-dichloroaniline mineralization by aniline. Appl Environ Microbiol 44:678–681.

You I-S, Bartha R (1982b) Metabolism of 3,4-dichloroaniline by *Pseudomonas putida*. J Agric Food Chem 30:274–277.

Young LY, Rivera MD (1985) Methanogenic degradation of four phenolic compounds. Water Res 19:1325–1332.

Zbozinek JV (1984) Environmental transformations of DPA, SOPP, benomyl and TBZ. Residue Reviews 92:114–155.

Zeyer J, Kearney PC (1982a) Microbial degradation of *para*-chloroaniline as sole carbon and nitrogen source. Pestic Biochem Physiol 17:215–223.

Zeyer J, Kearney PC (1982b) Microbial metabolism of propanil and 3,4-dichloroaniline. Pestic Biochem Physiol 17:224–231.

Zeyer J, Kearney PC (1983a) Microbial metabolism of [14C] nitroanilines to [14C] carbon dioxide. J Agric Food Chem 31:304–308.

Zeyer J, Kearney PC (1983b) Microbial dealkylation of trifluralin in pure culture. Pestic Biochem Physiol 20:10–18.

Zeyer J, Wasserfallen A, Timmis KN (1985) Microbial mineralization of ring-substituted anilines through an *ortho*-cleavage pathway. Appl Environ Microbiol 50:447–453.

Zeyer J, Kocher HP, Timmis KN (1986a) Influence of *para*-substituents on the oxidative metabolism of *o*-nitrophenols by *Pseudomonas putida* B2. Appl Environ Microbiol 52:334–339.

Zeyer J, Kuhn EP, Schwarzenbach RP (1986b) Rapid microbial mineralization of toluene and 1,3-dimethylbenzene in the absence of molecular oxygen. Appl Environ Microbiol 52:944–947.

Manuscript received April 6, 1988; accepted June 25, 1988.

Predicting Pesticide Residues to Reduce
Crop Contamination

E.I. Spynu*

Contents

I. Introduction

Man's efforts to restructure nature have led not only to unquestioned achievements but also to certain unforeseen negative consequences. A major example of this dichotomy is the widespread agricultural use of chemicals. Pesticides are widely used to control pests and plant diseases, which cause international losses amounting to $74.9 billion; this figure represents 50% of the entire cost of global agricultural production. Along with these important economic advantages, however, these chemicals produce significant environmental and human health effects.

The intensification of worldwide agricultural development has been accompanied by a marked increase in the spectrum of chemicals and the frequency of their use. This is shown in the fact that, while only 35,000 new chemicals were tested in the 15 years following the synthesis of DDT, approximately 10,000 per year have been produced since then. The high economic value of pesticides is also reflected in steady production increases. In 1964, for example, total world sales of agrochemicals amounted to $950 million; by 1971, international pesticide sales had risen to $2.32 billion. In the years 1975 to 1985, worldwide herbicide use increased from an estimated 2.3- to 4.54-fold; insecticides, 2.3- to 3.68-fold;

---

*Professor, All-Union Scientific Research Institute of Hygiene and Toxicology of Pesticides, Polymers, and Plastics; Ministry of Public Health, Kiev, USSR.

© 1989 Springer-Verlag New York Inc.
Reviews of Environmental Contamination and Toxicology, Vol. 109

and fungicides, 3.0- to 4.6-fold. In the USSR insecticide and fungicide use was
2.4 times greater in 1980 than in 1960, and herbicide use 14 times greater. Overall pesticide production is said to be between 300 and 400% greater (Vasiljev
1983; Melnikov et al. 1977; Melnikov 1979).

II. Health Effects of Pesticide Use

This substantial introduction of pesticides into the environment has inevitably
caused various human health consequences. The effects have differed, depending
on the strength and intensity of the exposure. Their widespread use over the last
20 yr has made available a great deal of data on different aspects of the adverse
human health effects of pesticides (Lazarev 1966; Medved 1971; Kundiev 1981;
Kagan 1984; Spynu 1965; Polchenko 1971). The pathogenic effects of pesticide
exposure are manifested in different ways. Highly toxic and/or cumulative pesticides can cause acute or chronic poisonings, alter immune responses, induce
allergic reactions, and produce mutagenic, teratogenic, or oncogenic effects.
Possible delayed physiological influences and pathological processes include a
number of reversible or slightly reversible effects such as atherosclerosis, hepato-
cirrhosis, pneumosclerosis, and demyelinization of nerve trunks. The notewor-
thy increase in allergic diseases is also assumed to be related to chemical
contamination of the environment.

From what is known about delayed pesticide reactions, of greatest concern are
carcinogenic effects. Minor tumorigenic effects have been identified from certain
organochlorine pesticides, such as DDT[1] aldrin, heptachlor, and methoxychlor.
Experiments with models have revealed the tumorigenic danger of a number of
dithiocarbamates: TMTD (thiram), zineb, ziram, and maneb. Malignant tumors
in different areas were detected in mice exposed to trichlorfon. One hundred
nineteen pesticides (49.8%) have shown mutagenic properties (Kurinniy and
Pilinskaya 1976; Kurinniy 1984; Pilinskaya 1984). Genetic activity has been
examined in only a limited number of substances, however; further investigation
would quite possibly reveal mutagenic properties in more pesticides. All of the
foregoing results were essentially achieved with models and not human subjects.

When the effects of zineb and ziram on the culture of human peripheral blood
leukocytes were studied, a marked increase in aberrative cells of the chro-
mosomal and chromatid type was detected. In addition, research that compared
leucocyte cultures from people who work with ziram with those of a control
group revealed significant changes. Definite parallels between human and animal
tests are noted in research on the cellular activity of maneb and TMTD.

Only a few years after the agricultural use of DDT and other stable organochlorine insecticides began, the potential for these substances to accumulate in the
human body became evident. Organochlorine substances were detected in the

---

[1] See Table 4 for chemical names of pesticides.

adipose tissue and blood of individuals who had never had any occupational contact with pesticides. These substances included DDT, DDE, γ-HCH (lindane), dieldrin, camphechlor (toxaphene), chlordane, and other organochlorines.

Studies on pesticide accumulation in humans established definite relationships between DDT in breast milk and the increased frequency of complicated pregnancies and deliveries, premature births, and newborn infants with lower birth weights (Medved 1971; Komarova 1983). Children's adipose tissue has been found to contain higher amounts of DDT and DDE than that of adults, a finding the author believed to be connected with the predominance of milk products and fats in children's diets (Vaskovskaya 1985). Some scientists have observed higher levels of P,P'-DDT in the blood of premature infants compared with full-term babies. On the other hand, certain researchers have found no positive correlation between human DDT content and any type of pathology (Maier-Bode 1966).

III. Predicting and Reducing Adverse Health/Environmental Effects of Pesticides

International research applies a complicated system for predicting the adverse effects of pesticides, especially those used in agriculture. Scientific studies are devoted to a large extent with establishing connections between the specific biological activity of a pesticide and its chemical properties. This research makes possible the selection of less toxic and persistent chemicals. Since the 1970s, because of study findings, a number of countries have prohibited agricultural use of DDT and certain other persistent chemicals. Many highly toxic organophosphate pesticides and some other classes of chemicals have also been banned. In the last decade, the overall toxicity of agrochemicals has decreased approximately 2–9 times. In addition, less stable pesticides are being substituted on a wide scale for the very persistent organochlorine chemicals.

Efforts to identify highly effective compounds that will eliminate all pests but spare any useful organisms in the environment are also widespread and continuing. Although a degree of success has been achieved in developing pesticides that protect nontarget organisms, increased knowledge about the delayed side effects of more and more chemicals is hindering this effort to develop pesticides that exhibit selective action.

Another avenue being pursued is the hygienic standardization of pesticides and establishment of action thresholds and tolerances. This concept is based on Gadamer's classic dogma of toxicology and pharmacology, "No substances are poisons in themselves; everything depends on the dosage." As evidence of the validity of this rule, highly toxic chemicals in small dosages are used for medicinal purposes. Standard tolerances can be established for most pesticides which result in food residues at levels below those believed to cause adverse effects, as well as ensuring that application rates do not cause environmental damage. Any chemical that cannot meet these requirements is cancelled for agricultural practices.

Determining residue levels and retention periods for pesticides, however, is not a simple task. International research for the past decade shows that in a number of countries overall pesticide levels in plant products are decreasing, especially for DDT and other banned chemicals. At the same time, DDT contamination of agricultural commodities remains sufficiently high to be a matter of concern. In Yugoslavia and Austria (Jarc 1980), DDT residues were detected in 100% of domestic animal adipose tissue samples. In 20 towns in India, high levels of DDT were detected in potatoes (0.01–1.0 mg/kg), eggs (0.047–0.97 mg/kg), milk (0.06–0.08 mg/kg), and wheat flour (0.1–10 mg/kg) (Kalra et al. 1978).

In France, high levels (16.0–19.5 mg/kg) of TMTD were detected in lettuce and spinach. In the German Democratic Republic (East Germany) and Italy, this pesticide was detected in cauliflower and lettuce (1.0–1.1 mg/kg), and apples and cabbage (0.03–0.08 mg/kg) (Ritely 1980). In the Federal Republic of Germany (West Germany), examinations of dill, borage, and other spices 35 to 60 d after treatment revealed that residues of 14 herbicides exceeded permissible levels. Residues of diene pesticides were found in hundredths of mg/kg (Norouha et al. 1980).

Data from Kuchak, Ivanova, and Voloschenko (1985) revealed that daily pesticide residue levels in people's diets in Poland, Hungary, England, Roumania, and the USSR vary in the range of hundredths to tenths of mg for DDT and TMTD and thousandths to tenths of mg for HCH (BHC), propyzamide (pronamide), and carbaryl. These data were generalized from the works of Liman et al. (1978); Smith (1978); Hoxton et al. (1979); Kaarchauskene (1973); Lootsoya et al. (1976); Shtenberg (1974); and Babaev et al. (1979).

Analysis of information about the comparative contamination from different points to the essential role of diet. The total annual intake of DDT and DDE by the human body amounts to approximately 50 mg. Of this, 44.8 mg comes from food, 0.03 mg from air, 0.01 mg from water, and 5.0 mg from contact with different commodities. Consequently, food contamination is receiving the greatest attention.

IV. Multiple Factors Affecting Residue Levels

Information about the levels of pesticide residues in agricultural commodities, however, can be obtained and assessed only at time of harvest. International assessments of pesticide residue levels are therefore always after the fact. If successful techniques are to be developed for decreasing residues, they must be based on knowledge of all factors affecting these quantities and must permit control before harvest. The many indices that characterize residue levels cannot always be strictly defined during application, vegetative periods, and harvesting. This inability to identify fully cause-and-effect relationships lowers the effectiveness of control measures in succeeding years. Effective control requires the development of quantitative and qualitative regulations that take into account the relationships between pesticide residues and the factors conditioning their levels. The literature shows the influence of different factors on residues in plants. Generalizations made from these data identify the following as the main influences: physical and chemical properties of the pesticide, application conditions (rates, restrictions, frequency, etc.), climatic characteristics of the environment, and chemical and structural peculiarities of treated sites. Compiling a bank of data on these influences would require a great deal of manpower, time, and funding; but currently available standards suffer due to lack of this information.

As an example, recommended standards governing post-treatment harvest waiting intervals do not consider all the necessary variations in existing conditions. If waiting intervals are shortened, treated products may be contaminated; if intervals are made longer, crops may be lost because of delayed harvesting. In addition, the fact that many years of testing are required to develop pesticides that act preventively complicates introduction of new, more effective products for plant protection.

V. Mathematical Modeling Techniques

If facilities could be established to permit the fast and accurate prediction of pesticide behavior under differing conditions without the need of years of actual use data, the process would be tremendously improved. Positive results would include development of adequate measures to eliminate contamination, establishment of optimal use standards, and faster introduction of new chemicals. The entire system of use, control, and treatment of pesticide exposures would operate more efficiently and with greater flexibility. The chief objective would be to find a fundamentally new and quick way to develop, through mathematical models, the best mix of measures to prevent adverse effects on human health and plants from greater pesticide use (Spynu and Ivanova 1970, 1972, 1974; Spynu 1979; Ivanova et al., 1975).

In pursuit of this objective, the following work has already been done:

1. Indices (factors) that characterize the process of pesticide degradation in the environment have been analyzed and selected. A significant base of experimental data on the dynamics of degradation under various combinations of specific environmental conditions and pesticides having different physical and chemical properties has been collected.
2. The validity of measurements of pesticide physical and chemical properties, degradation dynamics, and environmental factors has been assessed. The most probable hypotheses on the dependence of these dynamics on the values of related factors have been developed and examined. Exponential dependence is acknowledged as one of the most suitable theories; this ascertains the

decrease of pesticide contamination with the passage of time according to the time constant, determined by the chosen factors.

3. Existing mathematical methods of modeling multifactor relationships have been analyzed. Because of the need to evaluate many factors (more than 20) and the scarcity of experimental data, the method of group handling of arguments (MGHA) is considered suitable for accomplishing the task. This method is based on multiserial application of the Bayes formula and the formula of danger record, from the classical theory of statistical solutions.

4. To describe the initial concentration of pesticides on treated objects, an algorithm (MGHA) with sequential release of polynomial trends has been developed and applied.

5. To automate the laborious process of synthesizing and selecting mathematical models of pesticide degradation, specific computer programs using algorithms have been written. With these programs, certain mathematical models have been developed to describe the studied processes with sufficient accuracy for practical application.

6. Based on these models, instructions for predicting pesticide residues in agricultural products and treatment schedules (for vegetables and fruiting and berry-producing plants) have been prepared.

7. Theoretical and experimental results have been analyzed and compared with actual data on pesticide decomposition. The margin of error, determined to be in the range of plus or minus 20 to 30%, is deemed sufficiently accurate for practical use.

8. Specific standards of chemical use for agricultural plant protection in different zones of the USSR have been developed.

A. Factors Determining Variations in Pesticide Residues

The most significant advance has been the recognition of the dynamics of varying pesticide residues as a multiparameter process and the identification of the factors that determine these variations. Persistence of pesticides in the environment is considered the primary concern. This persistence depends on the physical and chemical properties of the specific substance and on other complex conditions, defined by various parameters. A specific chemical has been shown to remain in plants of different types for different periods. Analysis of organochlorine, organophosphates, and carbamate pesticides, applied to apples under similar conditions, shows that organochlorines exhibit the highest residue levels and longest periods of contamination during vegetation and harvesting. Carbamates are less persistent, and a number of organophosphates disappear quickly.

Studies on dimethoate, trichlorfon, fenitrothion, and fenchlorphos used under similar conditions on leaves of clover and potatoes reveal they dissipate over different periods of time. Pesticides have been shown to be very persistent on and in citrus fruit peel. Residue levels and rates of disappearance are basically affected by frequency of treatment and plant vegetative stages. As biomass increases at the beginning of vegetative periods, residues have been observed to dissipate intensively. The mode of pesticide interaction with plants, however, has not been sufficiently investigated.

When applied to plant surfaces, all pesticides penetrate tissues to a certain depth, which is influenced simultaneously by intraplant and external factors (such as photochemical transformation, volatilization, etc.). Many types of biotransformation are possible, including oxidation, hydroxylation, epoxidation, *N*-dialkylation, dechlorination, and reduction, and conjugation with sugars and amino acids. Indications that pesticides change to form complex compounds with plant substances have been observed. Organophosphate compounds and carbamates exhibit an especially high capacity to form complexes with carbohydrates and amino acids (Doterman 1972; Knaak 1972).

As they penetrate plant tissues, organic substances either transform to hydrophilic fractions or deposit in lipophilic tissues. Hydrolysis is seen as playing the most important role in chemical transformations; the rate at which these changes occur depends primarily on pH, temperature of the medium, and other factors. A number of organochlorine substances, because they are more highly soluble in fats than water, decrease very slowly and may deposit in lipophilic fractions. The reverse is true of many carbamates and organophosphate pesticides, accounting for their shorter persistence in plants.

Assessing the quantities of most enzymes that are present in plants and that participate in pesticide decomposition and transformation is not possible at present. At the same time, however, the indices of pesticide content appear to be closely interrelated with enzyme activity on other chemically active substances in plants. Identifying on a scientific basis the parameters that characterize the mode of pesticide/plant interaction is also currently not feasible. With this in mind, we have decided to record the maximum number of factors that can characterize plant chemical content.

Application conditions, such as methods, formulations, and rates, have a significant influence on pesticide levels and retention periods in treated plants. A number of pesticides have been observed to exhibit higher residues in plants with increased applications. Many authors consider the frequency of applications to be a significant factor in pesticide levels at harvesting and rate of disappearance during the vegetative period. In addition, a number of chemical transformations of pesticide structures are believed to be induced by light, temperature, humidity, and other climatic factors.

B. Defining the Role and Importance of Different Factors

From all this, the interaction processes of pesticides and plants can be seen to represent a complex system of dynamics. The measurable inputs of the system are the physical and chemical properties of a pesticide, chemical–structural peculiarities of plants, treatment conditions, and climatic factors. The result is the pesticide concentration found in an agricultural commodity at a given time. Preventing pesticide contamination of food requires regulation of this complex system of interrelated factors, even though the tasks of defining the role of each factor in this system and agreeing on its importance present considerable difficulty.

One problem is choosing the number of actually visible and controllable parameters. Certain essential features that define pesticide fate in plants have been completely overlooked, while very little information has been collected on others. Some factors play only indirect roles; some have qualitative but not quantitative meaning; and some are generalized concepts which still must be refined. The most significant problem, however, is the lack of information on the dynamics of specific pesticide residues in plants and how they correlate with the various factors that influence these processes.

A useful solution to such problems, when little is known about their composition, is the idea of a "black box," or a system comprised of a definite number of inlets and outlets with unrecognized internal modes of action. As new data about these internal modes of action and their relationships are compiled, "white boxes" composed of known components are substituted for the black boxes. The solution is thus based on clarifying the relationship between the outlet value and the factors that define it.

The world literature provides much information on the problem at hand, but the major portion of this information is useless because it is incomplete (i.e., dynamics with insufficient indices or a lack of fixed values for a number of factors). Additional data were collected to substantiate the definitive factors and the methods for developing the mathematical descriptions of the dynamics of plant pesticide content. A special database was developed to contain all the information on pesticide degradation in different plants and the determining factors; the contents included data on 90 processes of disappearance for 20 pesticides in 15 types of fruits and vegetables, under 14 sets of modified use conditions, in several different climatic-geographic zones of the USSR. The next step consisted of choosing a modeling method adequate for the inlet factors and selecting the most significant variables and combinations from the entire complex of information.

C. Use of Algorithms for Prediction

Because of the multiple factors and scanty a priori information, the use of determinant methods of approximation were not considered appropriate. As previously stated, the MGHA algorithms described by Ivahnenko (1975, 1982, 1983) and Ivahnenko and Osipenko (1982) offer many advantages because of their hierarchical structure. Using this method, a mathematical model of pesticide decomposition over time was developed. Model parameters are as follows:

$$C(t) = F(x, t), \tag{1}$$

where $C(t)$ is the current concentration of a substance in plants; t is the current time; and x is the vector of factors, defining the pesticide decomposition process.

Pesticide decomposition in plants may be viewed as a process in which the changing rate of current concentration (the first derivative of the concentration) is proportional to the pesticide quantity in a medium at a given moment in time:

$$\frac{dc\ (t)}{dt} = -MC(t) \tag{2}$$

where M is the coefficient of proportionality, named the coefficient of natural decomposition

$$M = +1/\tau,$$

where τ is the time constant of disappearance

The expression (2) is a common differential equation with separative variables. After integrating the expression (1), following the initial conditions, where $C(0) = C_0$, the following formula is derived:

$$C(t) = C_0 \exp\ (-t/\tau)$$

where C_0 is the initial concentration and τ is the time constant of pesticide disappearance.

A great number of test data, obtained by the least-squares method, show that at constant values of the parameter x the current concentration $C(t)$ for each pesticide is very closely expressed by this equation.

The principal hypothesis is therefore accepted, which states the dependence (1) refers to the following type of functions:

$$C(t) = C_0(x) \cdot \exp\ \{-t/\tau\ (x)\} \tag{3}$$

where $C_0(x)$ is the initial pesticide concentration in plants, depending on a number of factors x, and τ is the time constant of pesticide disappearance.

The vector of factors x defining the process of pesticide disappearance is described by the following components:

Physical and chemical pesticide properties:
x_1: molecular weight
x_2: melting temperature
x_3: solubility in fats
x_4: solubility in water
x_5: stability at pH 5 to 8
x_6: volatility
x_{17}: stability in alkaline media
x_{18}: stability in acid media

Chemical content of above-ground parts of plants:
x_7: pH of cell liquid
x_8: water content
x_9: fat content
x_{10}: protein content
x_{11}: total sugar content
x_{18}: total acidity
x_{19}: cellulose content
x_{20}: ash content
x_{21}: content of nitrogenous substances
x_{22}: content of pectin substances

Treatment conditions
x_{12}: frequency of plant treatment with a pesticide
x_{13}: rate of pesticide application per treatment

Climatic conditions
x_{14}: mean air temperature during period from last treatment until harvest
x_{15}: relative air humidity on day of treatment.

The initial concentration of $C_0(x)$ and time constant $\tau(x)$ may be expressed in different types of functions, including polynoms.

The basic principles of constructing the algorithms (MGHA) and descriptions of those that provided the most adequate models of the pesticide disappearance process are as follows:

1. Group handling of arguments represents a method of constructing a mathematical modeling system based on evristic self-organization, which permits progressive increases in the complexity of the model's mathematical description to increase its accuracy.
2. Gradual increases in complexity are controlled by the criterion value, such as the root-mean-square error, measured by individual testing of the initial data. This criterion minimum defines the model of optimal complexity.

To accomplish our objective, stoachastic algorithms (MGHA) were developed, based on "freedom of choice of decisions" (Gabor). This principle provides for a multiserial structured recognition system that generates many decisions for each series; with each succeeding series, any decision can be chosen from a preceding one. The appropriateness of earlier decisions therefore influences the effectiveness of later choices.

To determine $C_0(x)$ in the stochastic statement of the problem, the algorithm (MGHA) of the synthesis of polynomial mathematics is used, either in a multiplicative form or as the sum of power trends of initial variables. The algorithm is intended to identify the statistical and dynamic characteristics of complex objects from a small amount of experimental data. The complete description is expressed in a polynomial equation of regression. According to the algorithm for complex systems, the model $C_0 = F(x)$ is derived in this form:

$$C_0(x) = x_{13}/(0.94+0.012x_{13}+0.0062x_{15}) - (0.0143x_{14}+0.0008x_1+0.01x_6)/ \tag{4}$$

where x_1, x_6, x_{13}, x_{14}, and x_{15} are values of the previously described characteristics.

To define $\tau(x) = F_2(x)$ by the stochastic algorithms (MGHA), these values are obtained from the equation systems and confirmed with the vectors of the "x" characteristics, then divided into the following five classes:

Class I (R_1) $1 < \tau < 3.6$ d
Class II (R_2) $3.6 < \tau < 6.75$ d
Class III (R_3) $6.75 < \tau < 10.6$ d
Class IV (R_4) $10.6 < \tau < 15$ d
Class V (R_5) $15 < \tau < 30$ d

These class limits are set so that each should have about 20% of the patterns (observations) in the initial sampling. In terms of pattern recognition, the formula for approximating the time constant τ of the exponential of pesticide disappearance is as follows: the given values of characteristics x_1 through x_{22} enable the specific pesticide decomposition process to be assigned to one of the five classes R_i ($i = 1 \div 5$) by the value τ.

VI. Comparisons of Results

The stochastic algorithm (MGHA) model was compared with results obtained by other methods (regressive analysis, potential fractions, etc.) to ascertain its adequacy. The comparison used different criteria (xu − square, method of ranges, signs criterion and so on). When findings were compared, the algorithm models were seen to be superior to the analogous types.

The most reliable measure of adequacy, however, is the accuracy of agreement between the results of modeling and actual processes. With this in mind, extensive practical examinations of the estimated values of different standards of pesticide use were conducted. The dynamics of pesticide residues in plants were also studied in 14 different climatic areas of our country. These investigations produced data on the contamination of agricultural commodities in a unified form that would permit comparisons.

Two hundred dynamics (i.e., measurements of pesticide content in a plant at different periods after treatment) were studied. In all, 2,800 analyses were performed (if one dynamic is considered to include 6 min under duplicate or triplicate repetitions of study). To represent the main chemical classes (organophosphates, organochlorines, derivatives of thio- and dithiocarbamic acid) 11 pesticides were analyzed: dimethoate, methyl parathion, zineb, malathion,

Table 1. Comparison of actual and predicted pesticide degradation periods

		Class		
I	II	III	IV	V
$\dfrac{(18-15)\cdot 100\%}{18}$	$\dfrac{(18-21)\cdot 100\%}{18}$	$\dfrac{(18-20)\cdot 100\%}{18}$	$\dfrac{(18-20)\cdot 100\%}{18}$	$\dfrac{(18-14)\cdot 100\%}{18}$
$= 17\%$	$= 17\%$	$= 11\%$	$= 11\%$	$= 22\%$

trichlorfon, formothion, dicofol, carbaryl, phosalone, polychlorcamphene (toxaphene), and polychlorpinene (Strobane). Eleven types of fruits and vegetables were tested: apples, pears, plums, cherries, cabbage, potatoes, cucumbers, onions, corn, grapes, and tangerines.

Because our cooperative efforts were aimed at defining the most suitable times for harvesting treated crops, the data on the duration of pesticide retention in plants were of primary importance. The length of the process was estimated according to the classifications by the stochastic algorithms (MGHA). The correlation error of actual and estimated duration of pesticide retention in a treated site is expressed in the following formula:

$$\delta = \left(\frac{N - N_1(r)}{N}\right) \cdot 100\% \tag{5}$$

where $N_1(r)$ is the number of exactly classified processes in a given class. N is the number of processes, attributed to a definite class by the algorithm of classification.

Analysis of 90 processes of pesticide decomposition in various agricultural plants, divided by the classification algorithm into five classes (18 processes in each class) provided the accuracy shown in Table 1. As can be seen, the mean error in distinguishing the total duration of pesticide residue retention in plants was 16%; this is quite acceptable (from the expert point of view) for preventing adverse health effects from pesticide use and protecting plants. It will also permit using this process to establish estimated post-treatment harvest waiting intervals.

Table 2 presents data for certain pesticides comparing the actual and estimated retention times. The mean error, defined according to formula (5), is on the average 20% (Naldadjan et al. 1977; Ivanova 1975; Ordgzrnikidze et al., 1976). The formulas for predicting pesticide levels were examined in the same way, and the error between actual and estimated residue levels was calculated with the following:

$$\frac{C_{ac} - C_{es}}{C_{ac}} \cdot 100\% \tag{6}$$

Table 2. Comparison of estimated and actual pesticide retention times
in fruits in various regions

| | Retention time in days (actual/estimated) | | | | Mean error (%) |
Pesticide	Cherries	Sweet cherries	Apples	Tangerines	
Phosalone	28/30	30/32	30/35	22/21	4.5
Dimethoate	20/25	21/24	20/26	12/20	24.0
Fenitrothion	15/16	11/16	–	–	18.6
Formothion	–	–	18/25	18/25	28.0
Fosmet	–	–	20/26	12/21	33.0
Malathion	16/16	15/16	18/25	17/21	12.0
Phenthoate	20/25	20/25	–	–	–

where C_{ac} is actual content (mg/kg) of a pesticide at given moment of time and C_{es} is estimated content (mg/kg) of substance under defined conditions and at definite time after last treatment.

Results of these comparisons of predicted residue levels with experimental data from different climatic regions are presented in Table 3. Data in the table show that the error is not more than 20% when the contamination level does not exceed tenths of mg/kg, which is the level specific to identification of pesticide residues. If hundreds to thousandths of mg/kg are considered, the estimated values may be a great deal higher. The analytical method of inadequate sensitivity, however, also may produce deviations of the same magnitude in determining residue levels. Thus, the method of estimation is shown to provide adequate accuracy in practical use.

Our models of pesticide residue disappearance in treated plants are definitely of theoretical interest, since they permit identification of the relative importance in the process of different factors such as physical and chemical properties of pesticides, chemical composition of plants, treatment conditions, and climatic parameters. While 22 possible characteristics were identified as having possible effects on pesticide decomposition and transformation, the mathematical models show that 13 of these provide the most important information. The research results confirm the multifactor relationship of the processes, including the interrelationship of pesticide physical and chemical properties and the interaction and interchangeability of physiologically active plant components.

VII. Establishing Standards

Since the rate of pesticide degradation depends on multiple factors, the standards for pesticide use (such as waiting intervals, application rates, and frequency of treatments) should be based on the same number of considerations. Establishing

 E.I. Spynu

Table 3. Comparison of experimental data on pesticide residues with estimated values

Variations	Pesticide, crop and application rate	Day 1 (actual/ estimated)	Day 3 (actual/ estimated)	Days 5 & 6 (actual/ estimated)	Day 10 (actual/ estimated)
1	2	3	4	5	6
1	Dimethoate 3.6 kg/ha apples	0.8/1.05	0.8/0.73	0.4/0.33	0.4/0.2
2	Dimethoate 4 kg/ha apples	0.3/0.6	0.1/0.25	0.15/0.17	—
3	Dimethoate 4 kg/ha pears	−0.4/0.6	0.05/0.25	0.01/0.08	—
4	Zineb 3.7 kg/ha apples	1.3/1.9	—	1.3/1.5	0.68/0.8
5	Zineb 4.0 kg/ha onion	1.56/2.0	1.0/1.1	0.01/0.06	—
6	Phosalone 4 kg/ha apples	2.0/1.6	1.24/1.35	—	1.06/1.0
7	Phosalone 1 kg/ha apples	0.2/0.5	0.14/0.24	—	0.01/0.06
8	Trichlorfon 1.5 kg/ha apples	0.98/0.9	0.58/0.45	0.5/0.3	0.08/0.1
9	Trichlorfon 1.2 kg/ha apples	0.15/0.2	0.05/0.09	0.01/0.03	—

these regulations through experimental use, however, requires a great deal of manpower, expense, and time. Agricultural applications also call for thousands of standards because of the large number of pesticides, plants, and climate zones. Using the methods described, however, we have been able to develop safe pesticide use standards for all the parameters involved.

One of the most important areas for standards is the waiting interval. We have developed different schedules for various regions in the Ukraine; for this purpose, the entire USSR was divided into four zones based mainly on temperature/ humidity indices (sum of positive temperatures, hydrothermal coefficient, etc.):

Zone I: the wooded district and Precarpathian area
Zone II: the forest/steppe
Zone III: the steppe
Zone IV: the Transcarpathian area

Using the prediction formulas, we estimated the regularity of decreases in pesticide retention duration in plants in the following order: wooded district, Transcarpathian region, forest/steppe, and steppe. Accordingly, waiting intervals may vary as much as 10 d in different zones, even when pesticides are applied under otherwise equal conditions; for instance, 30 d in the wooded district after pesticide treatment until harvesting but 20 d in the steppe. We also lengthened the waiting intervals for a number of dangerous pesticides in certain zones, to reduce contamination of foods. The standard was also shortened for substances that are comparatively less persistent and toxic in a number of regions of the republic. These standards were expertly established through estimations that took into account data from the world literature on the process of pesticide disappearance in different commodities.

The major public benefit of this work has been the estimation and establishment of the principal standards for health and environmental protection before new pesticides are introduced into agricultural practice. These methods permit the prediction of dangerous contamination as well as the adverse effects of pesticide residues on the nutritional value of food products. In addition, they allow to establish pesticides use regulations based on computerized models, i.e. to prevent the adverse effects of pesticides on human health.

Summary

The system of dynamics between pesticides and plants is reviewed, and a conceptual model capable of reflecting the necessary qualitative and structural peculiarities is proposed as a means of predicting residue levels. The degradation processes of various chemical classes of pesticides in plants under different conditions of use are analyzed. Formulas are developed that enable recognition and estimation of residue levels and duration of retention for "new" pesticides and "old" substances under varying treatment conditions. Estimated data are verified to provide positive assessments of the accuracy of the predictions.

Mathematical modeling as a means of perception is stressed. With this method, the outlet value can be controlled by changing such inlet parameters as application rate, frequency of treatments, types of plants, and so on. Residue levels are predicted for different combinations of use conditions in various climatic-geographical regions.

The method of estimation also enables the development of important standards such as post-treatment waiting intervals. A more flexible technique can be

Table 4. Chemical names of pesticides mentioned in text[a]

Aldrin	1,2,3,4,10,10-Hexachloro-1,4,4*a*,5,8,8*a*-hexahydro-1,4-endoexo-5,8-dimethanonaphthalene
Camphechlor	(toxaphene): Reaction mixture of chlorinated camphenes containing 67–69% chlorine
Carbaryl	1-Naphthyl-*N*-methyl carbamate
Chlordane	1,2,4,5,6,7,8,8-Octachloro-3*a*,4,7,7*a*-tetrahydro-4,7-methanoindan
p,p′-DDE	1,1-Dichloro-2,2-bis-(*p*-chlorophenyl)-ethylene
p,p′-DDT	1,1,1-Trichloro-2,2-bis-(4-chlorophenyl)-ethane
Dieldrin	1,2,3,4,10,10-Hexachloro-6,7-epoxy-1,4,4*a*,5,6,7,8,8*a*-octahydroexo-1,4-endo-5,8-dimethanonaphthalene
Dicofol	(Kelthane): 4-Chloro-*a*-(4-chlorophenyl-*a*-(trichloromethyl)benzenemethanol
Dimethoate	*O,O*-Dimethyl-*S-N*-methyl carbamoyl methyl phosphorodithioate
Fenchlorphos	*O,O*-Dimethyl *O*-(2,4,5-trichlorophenyl) phosphorothioate
Fenitrothion	Phosphorothioic acid, *O,O*-dimethyl *O*-4-nitro-*m*-tolyl ester
Formothion	*S*-[2-(Formylmethylamino)-2-oxoethyl] *O,O*-dimethyl phosphorodithioate
γ-HCH	(Lindane) 1,2,3,4,5,6-Hexachlorocyclohexane
HCH	(BHC) 1,2,3,4,5,6-Hexachlorocyclohexane
Heptachlor	1,4,5,6,7,8,8-Heptachloro-3*a*,4,7-7*a*-tetrahydro-4,7-endomethanoindene
Malathion	*S*-(1,2-Dicarboxymethyl)-*O,O*-dimethyl phosphorodithioate
Maneb	[1,2-Ethanediylbis[carbamodithioato](2-)]manganese
Methoxychlor	1,1′-(2,2,2-Trichloroethylidene)bis[4-methoxybenzene]
Methyl parathion	*O,O*-Dimethyl *O*-(4-nitrophenyl)phosphorothioate
Phenthoate	ethyl *a*-[(dimethoxyphosphinothioyl)thio]benzeneacetate
Phosalone	*S*-[(6-Chloro-2-oxo-3-(2*H*)-benzoxazolyl)methyl] *O,O*-diethyl phosphorodithioate
Phosmet (fosmet)	*S*[(1,3-Dihydro-1,1-dioxo-2*H*-isoindol-2-yl)methyl]*O,O*-dimethyl phosphorodithioate
Polychlorcamphene	see Camphechlor (toxaphene)
Polychlorpinene	(Strobane): chlorinated mixed terpenes
Propyzamide	(Pronamide): 3,5-dichloro-*N*-(1,1-dimethyl-2-propynyl)benzamide
TMTD	(Thiram): Tetramethylthioperoxydicarbonic diamide
Trichlorfon	*O,O*-Dimethyl(1-hydroxy-2,2,2-trichloroethyl) phosphonate
Zineb	[[1,2-Ethanediylbis[carbamodithioato]](2-)]zinc
Ziram	(T-4)-bis(Dimethyldithiocarbamato-*S,S′*)zinc

[a]Chemical Abstracts Service (CAS) nomenclature.

employed, in which specific periods are established for different plants under various treatment conditions. Thus, both unjustified shortened waiting intervals or unnecessarily elongated intervals periods can be avoided.

The greatest value of the modeling approach is that information can be obtained on the degree of potential food contamination and major standards can be developed without the need for extensive experimental use of pesticides under actual conditions. This technique fully considers optimal use conditions for agrochemicals in terms of human health requirements and protection of plants. Mathematical modeling protects the environment while enabling the speedy selection of safe parameters for pesticide use conditions and realization of significant savings in manpower and time. Overall, this estimation method appears to be an efficient link in the broad system of preventing environmental pesticide contamination and protecting human health.

References

Babaev DA, Atzutskaya CG, Tarusova JuP (1979) Residues of organomercuric compounds in daily nutritional rations and in biomass. In: Transactions of Azerbaijan Scientific Research Institute of Virusology, Microbiology, and Hygiene, 24:44–47.

Doterman V (1972) Biological and nonbiological transformations of organophosphorous compounds: WHO Bull 1:135.

Hoxton J, Lindsay DG, Hislop VS, Solinon L, Dixon EJ, Evans WH, Reid JR, Henitl CI, Jeffries DT (1979) Duplicate diet study on fishing communities in the United Kingdom: mercury exposure in a "critical group". Environ Res 18:351–368.

Ivahnenko AG (1975) Long-term prognostication and control of complex systems. Technica Kiev.

Ivahnenko AG, Krotov GJ, Jurachkovsky JuP (1981) Exponential-garmonic algorythm MGHA. The Automatics 2:23–30.

Ivahnenko AG (1982) Cybernetics is the science of modelling links and control in complex systems. The Automatics 1:3–11.

Ivahnenko AG, Osipenko VV (1982) Algorythm of inverse conversion of stochastic characteristics in determined prognosis. The Automatics 2:7–15.

Ivahnenko AG (1983) MGHA features realizable in the algorythm of double-level long-term quantitative prognosis. The Automatics 2:3–17.

Ivanova LN, Russu-Lupan IT, Belaya AL, Matusevich TI (1975) On verification of prognosis of pesticide dynamics in plant products. Hygiene and Sanitary 6:105–107.

Jarc H (1980) Rückstandsuntersuchungen in üsterreich. 3 Mitt Untersuchungen über Schwermetall-, Pestizid-und Antibiotikarückstände in Schlachttieren aus Niederösterreich. Wien tierärztl Monatsschr 67(4):133–136, 138.

Kaarchauskene MP, Kenstarurene JuJu, Atmenskas JuI (1973) DDT and sevin residues in food-stuffs in the regions of Kaunas zone. In: Problems of Epidemiology and Hygiene in Lithuanian SSR, Gintaras, Vilnius, pp 213–214.

Kagan JuS (1984) Global importance of pesticides and peculiarities of their biological action. Handbook of prophylactic Toxicology, Moscow 2(1):123–134.

Kalra RL, Chawla RR, Dhaliwal GS, Joia SS (1978) DDT and HCH residues in foodstuffs in the Pinjab. Pros Symp Nucl Tech Stud Metabolism Effects and Degrad Pestic Tirupati, s.i.,s.a. pp. 19–30.

Knaak J (1972) Biological and nonbiological transformation of carbamates. WHO Bull 1:35.

Komarova LI (1983) Problem of pesticide burden and their bearing on development of some pathological processes. Use Hygiene, Toxicology of Pesticides and Polymers. Kiev Issue 13:177–182.

Kuchak JuA, Ivanova LN, Voloschenko ZA (1985) Automatic system of control of pesticide content in the environment. Zdorovje, Kiev, p. 79.

Kurinniy AI, Pilinskaya MA (1976) Investigation on pesticides as mutagens in the environment Naucova Dumka Kiev.

Kurinniy AI (1984) Pesticides as mutagens in the environment. Test-systems for assessment of mutagenic potential in contamination of the environment. In: Symposium, Irkutsk, pp. 25–26.

Lazarev NV (1966) Introduction in geohygiene. Nauka, Moscow.

Liman J, Piechocka J, Cwiertuiewska E, Wachnik Z (1978) Badania stopnia skazenia posilkow w sakladach zywieni zbirowego pozostalosciam i pestycydow. Rocz Paust Zakl Hig,9(5):491–498.

Lootsoya HI, Ilmoya NA, Uustanu LV, Sillaots HP (1976). Organochlorine pesticides content in man's food, in the environment and their burden in a man's organism under condition of stopping their use in agriculture. Actual Problems of Hygiene of Nutrition and Water. Tartu, pp 49–53.

Maier-Bode H (1966) Pesticide Residues. Translation from German. Mir, Moscow.

Medved LI (1971) Success and problems of pesticide use hygiene. Use Hygiene, Pesticides Toxicology, and Clinical Picture of Poisonings. Kiev, Issue 9:5.

Medved LI, Kundiev JuI (1981) Hygiene of the environment in connection with wide use of chemicals in agriculture. Vestnik Acad Med Sci USSR, 11:62–67.

Melnikov NN, Volcov AI, Korotkova OA (1977) Pesticides and the environment. Chemistry, Moscow, p 239.

Melnikov NN (1979) Perspectives of production and use of pesticides. Chemistry in Agriculture, 6:13–19.

Nalbadjan PA, Ivanova LN (1973) To description of dynamics of some pesticides in plant objects. Transactions of the VI Session of the Transcaucasian Council of Plant Protection, Tbilisi, pp 499–501.

Norouha A, Khandekar SG, Banerji SA (1980) Survey of organochlorine pesticide residues obtained from Bombey markets, and fields, and its hinterland. Indian J Geol 7:165–170.

Ordgzeninikidze EK, Ivanova LN (1976) On possibility and accuracy of estimation of the length-time of pesticide retention in products of fruitgrowing. Transactions of Scientific Institute of Plant Protection. Georgian SSR. Tbilisi. Issue XII:257–260.

Pilinskaya MA (1984) Genetic aspects of hygienic regulations of pesticides. Transactions of Union Congress of medical genetical workers. Moscow, p 259.

Polchenko VI (1971) Types of accidental poisonings with pesticides, their hygienic characteristics and classification. In: Use Hygiene, Toxicology of Pesticides, and Clinical Picture of Poisonings. Kiev, p 15.

Riteesy G, Braun HL, Frank R, Ripley BD (1980) Foliar pesticide residues in relation to worker reentry. Pestic Sci 2:643–650.

Smith SWC (1978) Pesticide residues in the total diet. Food Technol Austral 30(9):349–352.

Shtenberg AI (1973) Pesticide residues in foodstuffs. Medicine. Moscow.

Spynu EI (1965) Dependence of biological action of diene pesticides upon their chemical structure. Transactions of the IX Mendeleev Congress on General and Applied Chemistry. Nauka, Moscow. p 260.

Spynu EI, Ivanova LN (1970) Application of cybernetics approaches for solving problems of pesticide hygiene and toxicology. Hygiene and Sanitary, 10:50–52.

Spynu EI, Ivanova LN (1972) Mathematical prognostication of the length-time of pesticide destruction in plants by stochastic algorythms of the method of group handling of arguments. Hygiene and Sanitary, 10:43–48.

Spynu EI, Ivanova LN (1974) Modelling-prognostication-regulation-in the system of prophylactic measures against environmental contamination. Vestnic Acad Sci UkSSR. Kiev, 2:78–87.

Spynu EI (1979) Mathematical modelling and regulation is efficient way in protection of the environment from pesticide contamination. Transactions of Soviet-American Symposium, Tbilisi, pp 128–133.

Manuscript received July 7, 1988; accepted August 6, 1988.

Association of Official Analytical Chemists: 1964–1988

Helen L. Reynolds*

Contents

I. Introduction

In 1964, William Horowitz published the history of the Association of Official Agricultural Chemists (AOAC), of which he was the chief executive officer (Horwitz 1964). At that time the AOAC had existed for 80 years and had ventured very little beyond its stated purpose of validating and publishing standardized methods of analysis for substances important to agriculture and the public health through a highly structured system of interlaboratory testing and review. In the ensuing quarter of a century, however, the AOAC not only changed its name to Association of Official *Analytical* Chemists but also underwent a striking expansion and transformation. The highlights of that transformation are the subject of this account.

In addition to structural changes, the AOAC greatly extended the range of its scientific activities. Because of space limitations, only a few areas of interest will be covered in detail, namely, pesticide residues, metals, and other elements in foods, and mycotoxins. Brief summaries will be given of some recent work on other topics.

---

*U.S. Food and Drug Administration, Department of Health and Human Services, Washington, D.C., retired. Past President and Fellow, Association of Official Analytical Chemists.

Reviews of Environmental Contamination and Toxicology, Vol. 109.

The Association celebrated its centennial in 1984, which included the publication of a history of AOAC by Helrich (1984). Much of the historical information in the present article is based on that publication.

II. Background

A brief history of AOAC is essential to an understanding of its present status. In 1884 a group of state chemists met in Philadelphia to attempt to solve one of their most pressing problems. They had the responsibility for certifying the contents of commercial fertilizers being marketed as a consequence of nineteenth-century advances in scientific agriculture. Despite their analytical competence, they were obtaining discrepant results and realized that the chief cause was the lack of standardized methodology. Having tried, and failed, to work through one of the existing scientific societies, they formed a new one, the Association of Official Agricultural Chemists. The word "official" clearly indicated the composition of the membership: these chemists were regulatory officials employed by some branch of government. The question of participation by industry had already been debated, and both industry and government had recognized that decisions on acceptable methods would be more authoritative if made by persons with no financial interest in the results. This principle has continued through its first century.

A. Administrative Structure

The administrative structure was relatively simple. The Association was administered by a Board of Directors consisting of a President, Vice President, Immediate Past President, and three additional members, one elected each year. It was tradition for an officer to remain an "additional member" for three years, then be elected Vice President the fourth year, President the fifth, and Immediate Past President the sixth year. A Secretary or Secretary-Treasurer was also elected and supervised the day-to-day operations of the Association. The Board usually met once a year at the AOAC annual meeting. Later, as activities of AOAC became more complex, the Board agreed to meet more frequently and to take a much more active part in Association matters.

The Secretary-Treasurer, though subject to yearly election, usually served for a number of years (Horwitz served for 25 years) and spent a considerable portion of his time on AOAC duties, with the approval of his employer (*see* the section on Government–AOAC Relationship). The President's duties were chiefly to preside at Board meetings and to approve appointments to Refereeships and committees; the position was essentially honorary. It was traditional for many years that the composition of the Board reflect participation in AOAC: United States and Canadian national government agencies, and state and provincial agencies. The methods adoption process was always separate from the administration of AOAC, and the Board of Directors had no authority in adoption or rejection of methods. This has not changed.

B. The Referee System and the Methods Adoption Process

The founders of AOAC worked out a pattern that is still followed in its essence. A specific topic is assigned to a scientist to investigate, e.g., a method to determine nitrogen in fertilizer. This scientist, or Associate Referee, surveys the literature and selects a promising method or develops a new method, which is then subjected to intralaboratory study. When the method appears to meet the criteria of accuracy, precision, reliability, and suitability for routine use, the Associate Referee arranges for a trial of the method by a number of analysts, or collaborators, working independently. Collaborators are provided samples or instructions on how to obtain them, and are given explicit written direction for the analysis and the reporting of results.

The Associate Referee receives, evaluates, and forwards the results, together with a recommendation as to whether the method should be accepted, rejected, or studied further, to the General Referee, an expert in a broader subject, e.g., fertilizers, who oversees the efforts of a number of Associate Referees. The General Referee examines reports of all Associate Referees within the major topic; decides on acceptance, rejection, or further study of all methods under consideration; and forwards a report containing these recommendations to a higher level (formerly the Committee on Recommendations of Referees), the Official Methods Board. This Board is composed of experts in fields of interest to the AOAC familiar with the AOAC's procedures and criteria for acceptance of analytical methods. The Board is divided into committees covering very broad areas (in our hypothetical case, the Committee on Feeds, Fertilizers, and Related Materials).

The appropriate committee of the Official Methods Board reviews the General Referee's report; decides on acceptance, rejection, or other disposition; and prepares a final recommendation to be submitted to voting members of the Association at its annual business meeting. Methods are usually adopted as official first action on the basis of the collaborative study; then, after a method has been in use for a few years so that any unforeseen problems have time to surface, it may be raised to official final action status. Both first action and final action methods are "official" and enjoy the endorsement of AOAC. If later evidence clearly indicates that a method is faulty it may be repealed by the same two-stage process: a vote for first action repeal one year, followed by a vote for final action repeal in a following year. The process has several possibilities for appeal at various stages. Details of adopted methods are published in "Official Methods of Analysis of the AOAC."

To help speed the adoption process, the concept of "interim" adoption was instituted in the late 1970s. A method for which the collaborative study was completed too late for consideration at the annual meeting but considerably in advance of the next meeting could be submitted during the interim, together with the supporting data, for review by the General Referee and the Official Methods Board. If the General Referee and the Official Methods Board approved the method it could be adopted as "interim first action" and could be used immediately as an official AOAC method. The formal vote for adoption would then take place at the following annual meeting. Acceptance of this innovation was slow at first, but is now well established, and the majority of methods are adopted first as interim methods, with a subsequent vote by the entire Association.

Another innovation was the introduction of the "surplus method" concept. It had become obvious that many methods remained in "Official Methods of Analysis" for years after they had ceased to be useful. The methods were still valid but superseded by newer, improved techniques. Most analysts simply bypassed these obsolete methods for more modern alternatives, thus they were considered as surplus, still official but no longer needed. A method being considered for surplus status was accordingly flagged with an asterisk in the next edition of "Official Methods." If five or more users indicated they still used the method, the asterisk was removed in the next edition and the method retained. If fewer or no laboratories expressed interest, the details of the procedure were omitted in the next edition, and the method was included by title alone, although it retained official status. The few potential users were referred to the last edition in which detailed Procedures appeared.

This methods adoption procedure is essentially conservative and has often been criticized as slow and cumbersome. Yet it has withstood the test of time. The various levels of review permit dissenting opinions to be heard. The time lags prevent premature adoption of an unreliable method. A method that has gone through the entire procedure generally proves to be usable until a better one is developed.

C. The AOAC–Government Relationship

To return to the early history of AOAC, one of the founders and an early officer was Harvey W. Wiley, the "Father of the Pure Food Law." Wiley had left his position as Indiana State Chemist and professor of chemistry at Purdue University, to accept a post as chief of the Division of Chemistry of the U.S. Department of Agriculture (USDA) at Washington, D.C. During his long tenure, he was active in AOAC and served over 20 years as its secretary. Because of his work at USDA, Wiley broadened the scope of AOAC to include work on foods, particularly contamination of foods by various substances, and on drugs, since at that time most drugs were plant extracts and thus were appropriate subjects of agricultural science.

The USDA acted as sponsor of the Association for many years, hosting its annual meeting, providing office space, secretarial assistance, and permitting USDA scientists to act as officers of the Association. For the first 30 years, the proceedings of AOAC meetings and the adopted methods were published as bulletins of the USDA, and were edited by USDA staff. The AOAC also played some part in the fight to enact the first Pure Food and Drugs Act.

When Wiley left the USDA in 1912, however, the Department decided that it would no longer publish the proceedings or the official methods, although continued other forms of support for years. The AOAC subsequently founded the Journal of AOAC, financed by the sale of subscriptions and that contained the proceedings of the annual meeting. In 1920 the Association began the publication of its compendium of methods, now known as "Official Methods of Analysis of the AOAC." The Book of Methods, as it is commonly known, immediately proved successful, and its sales still form a large part of AOAC's income. "Official Methods" is revised at approximately 5-yr intervals and is currently in its 14th edition.

After the Food and Drug Administration (FDA) was separated from the USDA and became an independent agency, it took over sponsorship of AOAC in much the same form as USDA had: FDA provided office space and telephones, and permitted one of its scientists to act as chief executive officer of AOAC. Other scientists were permitted and even encouraged to serve as officers, committee members and chairmen, Referees, and collaborators. The AOAC hired its own office staff (for many years only one full-time and one part-time person) and paid for postage, supplies, and publications. This arrangement lasted well into the 1970s. The FDA, and other federal and state agencies, recognized the value of having access to reliable and recognized methods of analysis and the efficiency of having an organization to oversee their testing and validation.

III. Transformation

The AOAC continued in much the same mode, with some relatively minor changes, from the time of Harvey Wiley's departure from government service until the 1970s. The number of referee topics investigated and the number of official methods adopted increased slowly but steadily. Whenever new legislation increased responsibilities of regulatory agencies responsible for agriculture and the public health, AOAC activities increased proportionately. The number of subcommittees comprising the Committee on Recommendations of Referees grew from four in the early 1950s to seven. Microbiological methods became a significant topic, with a large number of Associate Refereeships, and many microbiological methods were adopted.

Each edition of "Official Methods of Analysis" was larger than its predecessor, and the number of the Journal pages doubled. Early in its existence, the Journal's scope expanded beyond a simple record of the proceedings of the annual meeting to include papers on new developments in methodology. By the 1960s, a peer review system for manuscripts was in place, and the Journal operated essentially like other similar periodicals. AOAC undertook new publishing ventures in addition to "Official Methods" and the Journal: monographs, manuals, and full-length, hard-cover books. The Editorial Board expanded and met more often. Together with the Board of Directors and the Official Methods Board, it became a major force in AOAC governance.

To meet other needs, a number of standing and ad hoc committees were instituted, and contributed to AOAC work, i.e., The Long-Range Planning Committee, Committee on the Constitution of AOAC, Committee on Statistics, and Committee on International Cooperation. The annual meeting, traditionally held during October in Washington, D.C., was augmented by a spring training workshop held in different major cities. The full-time AOAC staff had increased, and the space provided by the FDA was becoming inadequate.

A. Changing Status

In the early 1970s, a committee appointed by the Commissioner of Food and Drugs to review FDA's management and operations recommended a change in the relationship between the FDA and the AOAC. The committee recommended that AOAC become completely independent, with its own premises and sources of support, and that its chief executive officer be an employee of AOAC, rather than a government scientist. This change eventually took place, although not for several years and not in a single action. Lengthy discussions ensued, and the AOAC did not move into its own premises until 1979.

Meanwhile an Executive Director had been hired and had completed an orientation period under Dr. William Horwitz of the FDA, who had served as the AOAC's chief executive officer since 1952. The by-laws had been revised to accommodate the changed status. New sources of financial support were sought, including membership dues, a mixture of contracts, grants, and agreements with various U.S. and Canadian government agencies, sustaining funds from state and provincial counterpart agencies, and private sustaining memberships from the various organizations served by the official methods process.

Several other changes had taken place that were not direct results of the transition but were related. The annual meeting was no longer restricted to Washington, D.C., but was scheduled for various other locations. The 1986, 1987, 1988, and 1989 meetings were held in Scottsdale, AZ, San Francisco, CA, Palm Beach, FL, and St. Louis, MO, respectively. The 1990 meeting is scheduled for New Orleans, LA.

In the early 1980s regional sections were instituted. The first charter was granted to the Pacific Northwest regional section. At present there are seven additional sections: Central, New York–New Jersey, Midwest, Southeast, and Northeast United States; Mid-Canada and Eastern Ontario–Quebec in Canada. Western Canada participates in the Pacific Northwest section. Regional sections carry on active programs and provide an opportunity for participation in AOAC by persons not able to attend the annual meeting. Since the sections hold local meetings and workshops, the Spring Training Workshop was dropped.

B. New Roles for Nonofficial Scientists

Scientists from industry and the academic sector had participated in methods development and testing since the early days of AOAC, but the by-laws precluded their taking part in the methods adoption process. Once the Association was no longer sponsored directly by a federal agency and industry was contributing substantially to its support, a movement arose to grant full membership to scientists other than regulatory officials.

The first step was to appoint members of industry and the academic sector to General Refereeships, the basic level in the methods adoption process. Later these scientists became eligible to serve on the committees that comprise the Official Methods Board. Finally, at the 1987 AOAC meeting, the voting members agreed to confer full membership, including voting rights, to all eligible scientists, regardless of their employment status, enabling them to serve as officers of the Association. In other organizations, this status would be taken for granted, but to the AOAC, after more than 100 years, in which only regulatory officials were eligible to accept or reject proposed methods of analysis, it was a step of the greatest significance. Too short a time has passed to determine whether this change will make any real difference in AOAC's effectiveness.

IV. The Collaborative Study

One aspect of the AOAC's structure not likely to change in the foreseeable future is the emphasis on the collaborative study. As Burke (1980) commented in his General Referee report, "The collaborative analytical method testing process is the foundation of the AOAC. It is a tool as important to the regulatory analytical chemist as are the instruments in his laboratory. It is our job as analytical chemists and managers to see that this instrument, the collaborative study, is properly used . . . to give (collaborative studies) top priority attention in terms of time, analyst expertise, and laboratory facilities. Anything less is a gross misuse of laboratory facilities and staff."

Actually the collaborative study system, described in Part II above, has undergone considerable strengthening since its inception, principally through the use of statistical methods in study design and results evaluation. Initially no minimum number of collaborators or samples was specified, and occasionally the collaborators were all from the same organization. As Helrich (1984) describes in his history of AOAC, the importance of proper statistical design of experiments and evaluation of results led to the formation of the Committee on Statistics and the presentation at the 1961 and 1962 meetings of several lectures by W.J. Youden, of the U.S. National Bureau of Standards, on statistics and their application to AOAC collaborative studies. These were so well received and Dr. Youden, a chemist turned statistician, was so skilled in presenting statistical principles to chemists, that he was asked to prepare statistical guidelines for the Association. This task eventually developed into a monograph (Youden 1967), which has been revised, reprinted, and later combined with a work by Steiner (Youden and Steiner 1975).

The Committee on Statistics recommended minimal collaborative test requirements for the adoption of methods in 1968. Later a consulting statistician was assigned to each of the committees comprising the Official Methods Board. Statistical tests are now used routinely by Associate Referees to evaluate methods and make recommendations for adoption.

The AOAC has participated in international efforts to harmonize collaborative studies. At the International Union of Pure and Applied Chemistry (IUPAC) Workshop on Harmonization of Collaborative Analytical Studies (Guidelines 1988), the following recommendations were accepted by consensus:

1. *Minimum number of materials*: five (only when a single level specification is involved for a single matrix may this minimum be reduced to three).
2. *Minimum number of laboratories*: eight reporting valid data for each material (only in special cases involving very expensive equipment or specialized laboratories may the study be conducted with an absolute minimum of five laboratories, with the resulting expansion in the confidence interval for the statistical estimates of the method characteristics).
3. *Minimum number of replicates*: one, if within-laboratory repeatability parameters are not desired; two, if these parameters are required. Replication should ordinarily be attained by blind replicates or split levels (Youden pairs).

The Guidelines also recommended that the final report contain a description of the materials used, their preparation, any unusual features in their distribution, and a table of all *valid* data, including outliers; that the report be published in a generally accessible publication or its availability indicated; and that it be sent to all participants.

The AOAC has developed a set of three videotapes covering the design and conduct of the collaborative study, use of statistics in evaluating the results, and preparation of the report for publication. It is a requirement for adoption of a method by AOAC that the report of the collaborative study, including tabulation of the results, be presented at the annual meeting and published in the AOAC Journal.

V. Advances in Methodology

The following is an account of some of AOAC's activities in development and validation of methods during the past 25 years. Because of space limitations, only three areas can be described: pesticide residues, metals, and mycotoxins in foods. Those three areas, however, were extremely active during that time. Unless otherwise stated, the collaborative studies cited, all resulted in adoption of the methods. Later revisions could not be included.

A. Pesticide Residues

During the early 1960s, pesticide residues were still being determined as individual compounds. Martin and Schwartzman (1964) conducted a collaborative study in which five laboratories determined nicotine residues in apples and spinach; one of the five also determined nicotine in cabbage. Nicotine was stripped from leafy crops with ammoniacal benzene-$CHCl_3$ and from waxy crops with NH_4OH, extracted with HCl, and read on an ultraviolet spectrophotometer. Pasarela (1964) studied a method for dodine (dodecylguanidine acetate) based on formation of a complex with bromocresol purple, followed by spectrophotometric determination at 590 μm. Seven collaborating laboratories determined total dodine residues in apples, four collaborators determined it in peaches, two in pears and strawberries, and one in pecans.

Five collaborators participated in a study of an unpublished method of Shell Development Company modified by Bluman (1964) for use with apples, tomatoes, and cabbage. Quantitative determination was based on cholinesterase inhibition and spectrophotometric measurement of the absorbance at 540 nm of the resulting red complex.

The emphasis soon changed to multiresidue methods, in which a number of pesticides could be simultaneously determined in the same sample. In his review of the history of pesticide residue analysis by AOAC methods, Bowman (1984) described the tremendous advances in methodology during the mid-1960s. The highly sensitive electron capture (EC) detector of Lovelock and Lipsky (1960) coupled with the use of gas chromatography (GC) for organochlorine (OCl) pesticide residue determination by Goodwin et al. (1961) and by Watts and Klein (1962) enabled detection at the picogram level. The extraction and cleanup systems of Mills (1961) and Johnson (1962) were combined with quantitative determination by gas-liquid chromatography (GLC) and confirmation by paper or thin layer chromatography (Kovacs 1963).

Johnson (1965) successfully applied the multiresidue system to the determination of OCl pesticide residues in butterfat and evaporated milk in a 20-laboratory collaborative study. This system was then extended to nonfatty vegetables by Krause (1966), leafy and cole-type vegetables by Gaul (1966), and small fruits by Davidson (1966). Further work centered on extending the multiresidue methodology to additional pesticides and crops. In his General Referee report, Burke (1968) summarized a study by six laboratories in which the method was validated for nine OCl pesticides on an additional 15 fruits and vegetables, bringing to more than 30 the total number of crops for which the multiresidue methodology could be used. Other collaborative studies involving the multiresidue technique for OCl pesticides included Carr's work on butterfat (1970) and fish (1971) that increased the number of pesticides detected, and that of Krause (1973), which extended the butterfat determination to include Perthane, heptachlor epoxide, mirex, and dieldrin in apples and cauliflower, as well as Perthane in butterfat.

Along somewhat different lines, Woolson (1973) conducted a collaborative study of the extraction of OCl insecticides from soil. Seven laboratories extracted the insecticides from three unknown soil samples containing six added insecticides each. The soils were premoistened with 0.2 M ammonium chloride by shaking with hexane–acetone; cleanup and determination procedures were those of the official AOAC multiresidue methods.

Other studies involved refinements of the methodology or extensions to difficult commodities. Burke (1980), in his General Referee report, described a macroscale saponification for the supplemental cleanup of the multiresidue method eluate of fatty foods. The microsystem accomplished the same result but required less eluate and reagents, took less time, was easier to use, and eliminated a subsequent step. Erney (1983) developed a rapid screening procedure which involved a miniaturized Florisil column, for OCl pesticides and polychlorinated biphenyls (PCBs) in fish. The final determination was based on GLC with EC detection, comparable to the official methodology. Seven collaborators analyzed 10 samples, with recoveries ranging from 79.3 to 100.7%. The method reduced reagent volumes by 80% and analytical time up to 50%.

Ault and Spurgeon (1984) extended the multiresidue concept to determination of OCl pesticides in poultry fat. The samples were rendered and cleaned up by automated gel permeation chromatography. Nine laboratories analyzed 10 samples fortified with 16 nonionic chlorinated hydrocarbons and one incurred residue sample, with an average recovery of 91.9%. Bong (1977) conducted a collaborative study of hexachlorobenzene and mirex in butterfat and fish by eight laboratories on samples of sablefish and fortified butterfat. Watts et al. (1980) studied an improved method (Crist et al. 1975) for hexachlorobenzene and mirex in adipose tissue in which an extracted or rendered fat sample was applied directly to a Florisil cleanup column, eluted in one fraction, and measured by direct GLC of the concentrated eluate. Twelve collaborators analyzed 10 samples of chicken fat.

In 1966 Giuffrida described the use of the alkali flame ionization detector with organophosphorus (OP) pesticide residues. In 1965 Storherr and Watts developed the sweep co-distillation cleanup method for this class of pesticides. Their cleanup used the "Storherr tube," 24 cm long by 5 to 6 mm inside diameter, which contained glass wool. The determinative step used GLC with detection by the thermionic detector of Giuffrida (1966). The technique could be applied to crop extracts of 2 g or less. Storherr and Watts (1965) applied the method to kale, carrots, apples, strawberries, and potatoes fortified at 1.0, 0.5, and 0.1 ppm with a mixture of Trithion, diazinon, malathion, methyl parathion, and parathion. Average recoveries ranged from 89 to 101%. The cleanup required about 20 min per sample and produced better results than absorption column chromatography. Watts and Storherr (1967) extended the technique to determination of both OP and OCl pesticides (electron capture was used for the OCl), and Storherr et al. (1967) applied it to 14 OCl and 10 OP residues in edible oils.

Wessel (1967) used the thermionic and EC detectors to extend the multiple detection GC method for determination of ronnel, ethion, Trithion, diazinon, methyl parathion, parathion, and malathion in leafy vegetables and fruits. Twelve collaborators from six laboratories obtained average recoveries for the seven residues of 88.7% by the thermionic detector and 91.2% by EC at the 0.5 ppm level from apples, and about 89% at the 5.0 ppm level from lettuce by both detectors. Finsterwalder (1976) conducted a collaborative study of an extension of the Mills method in which 15 laboratories determined parathion, o,p'-DDT, p,p'-DDT, and p,p'-DDE in kale, and lindane, dieldrin, heptachlor epoxide, and p,p'-DDE in eggs. The thermionic detector was used for parathion and EC GLC for the OCl pesticides.

In 1968 Storrher and Watts conducted a collaborative study of the sweep–co-distillation cleanup and GLC determination with the thermionic detector in which nine collaborators in five laboratories used the method to determine diazinon, methyl parathion, malathion, parathion, ethion, and carbophenothion in kale, carrots, lettuce, apples, potatoes, and strawberries, with over 90% average recovery. Laski (1974) conducted a collaborative study of this method for OP residues in apples and green beans in which 10 laboratories determined six OP pesticides or their oxidation products with the thermionic detector, and five of the laboratories also ran the method with the flame photometric detector in the phosphorus mode. The results warranted official first action adoption of the method for parathion, paraoxon, carbophenothion oxygen analog, carbophenothion, and EPN, but reagent interference made dimethoate quantitation inaccurate. The comparison study showed that the two detection systems were equivalent in precision and accuracy.

The most recent study of multiresidue methods was the collaborative study conducted by Sawyer (1985) in which 10 laboratories analyzed unfortified and fortified samples of lettuce, tomatoes, and strawberries for OCl and OP pesticides by applicable portions of the comprehensive multiresidue method of Luke et al. (1981). The three crops were fortified with alpha-BHC, dieldrin, chlorpyrifos, acephate, omethoate, and monocrotophos, each at three levels per crop. Sixteen pairs of blind duplicates were included in the 54 fortifications. The OCl pesticides were chromatographed on a methyl silicone column and detected with a Hall 700A electrolytic conductivity detector; the OP pesticides were determined with a flame photometric detector after chromatography on a DEGS column. Chlorpyrifos was quantitated on both systems. The procedure required essentially no cleanup before GC and proved comparable to existing multiresidue methods for pesticides of these types. The AOAC adopted the methodology with both GC systems.

Space permits only brief mention of methods for other classes of pesticides. Johnson (1964) conducted a collaborative study of the carbamate pesticide carbaryl in which five collaborators determined carbaryl residues in apples and lettuce by a method based on alkaline hydrolysis and colorimetric determination of the resulting 1-naphthol with *p*-nitrobenzenediazonium fluoborate as the chromogenic agent. Palmer and Benson (1968) reported a collaborative study of carbaryl residues in apples and spinach, in which eight laboratories tested a method involving extraction, cleanup, and semiquantitative thin layer chromatographic determination. Three FDA district laboratories also tested the method on 9 of the 12 commodities of the Total Diet with satisfactory results. Holden (1975) directed eight collaborating laboratories in a study of a multiresidue method using 2,4-dinitrophenyl ether for determining propoxur, carbofuran, carbanolate, and carbaryl in apples, corn, green beans, and leafy vegetables at three fortification levels. GLC was used for the determinative step. Certain precautions were necessary for carbaryl because of its instability in analysis. In 1985 Krause reported the successful adaptation of liquid chromatography to the determination of *N*-methylcarbamates and their metabolites in grapes and potatoes. Six laboratories used a method in which carbamate residues, after extraction and cleanup, were separated on a reversed phase liquid chromatographic column with an acetonitrile–water gradient mobile phase and the eluted residues were detected by an in-line post-column fluorometric detection technique. Seven carbamates and two metabolites were determined at two levels in blind duplicate samples.

Other methods were adopted for single sweep oscillographic polarographic confirmation of OP pesticide residues in nonfatty foods (Gajan 1969); ultraviolet determination of naphthalene acetic acid in apples and potatoes (Randall 1970); endosulfan and related products by EC GC (Mitchell 1976); ethylene thiourea in potatoes, spinach, applesauce, and milk by GLC (Onley 1977); and ethylene dibromide in grains and grain-based products (Sawyer and Walters 1986).

B. Metals in Foods

For some years, methods for metals and other elements in food were combined with those for pesticide residues under the topic, "Metals, Other Elements, and Residues in Foods," probably because most early pesticides were preparations of the heavy metals, lead and arsenic, or other elements (fluorine). With the widespread development and use of the organohalogen and OP pesticides, the topic was split. Arsenic, cadmium, lead, and mercury in various foods remained a matter for concern. Other toxic metals were also the subject of methods studies.

Bartlet (1964) reported on the Rhodamine B colorimetric method for antimony, studied by the American Conference of Governmental Industrial Hygienists. Ten collaborators analyzed three samples by a method in which a colored complex was formed between antimony and Rhodamine B in acid solution. Butrill (1973) conducted a collaborative study for determining arsenic residues in red meat and poultry by a colorimetric method used in the Meat Inspection Division of the USDA. The sample was ashed in the presence of magnesium nitrate at 600°C. The ash was dissolved in dilute HCl and zinc was added to generate AsH_3, which was trapped with iodine solution; the molybdenum blue color was developed with ammonium molybdate solution and read at 840 nm. Nine collaborators each analyzed six groups of four samples.

Ihnat (1974) reported a collaborative study of a fluorimetric method for selenium in which samples are subjected to modified wet digestion, first in HNO_3, then in HNO_3-$HClO_4$-H_2SO_4, and selenium is determined by reaction with 2,3-diaminonaphthalene and fluorometric measurement. Nineteen laboratories reported analyses of 10 samples representing vegetables, cereal and dairy products, meat, and fish, and containing naturally occurring selenium. Two National Bureau of Standards standard reference materials were also analyzed.

The introduction of atomic absorption spectrophotometry (AAS) gave considerable impetus to methods for metals in food, and a number of methods were adopted by AOAC. Rogers (1968) directed a study in which 10 collaborators used AAS to determine zinc at levels from 5 to 60 ppm in three sucrose solutions of known zinc content as well as soya meal, white flour, and whole wheat flour. Both wet and dry ashing were used as the preparatory step. Munns and Holland (1971) used a method for mercury in fish in which volatilized mercury vapor was measured by flameless AAS. Fish muscle was digested with a mixture of HNO_3, H_2SO_4, and $HClO_4$, and the mercury was reduced with stannous chloride–hydroxylamine and volatilized at room temperature in a stream of air. Nine laboratories obtained better recoveries than with the existing official method. Krinitz and Holak (1974) reported a rapid technique for mercury in seafood in which the sample is rapidly digested in a Teflon vessel using a closed system, and mercury is determined by flameless AAS as described by Munns and Holland (1971). Ten collaborating laboratories analyzed two samples each of shrimp and tuna and a gelatin mercury reference standard. Munns and Holland (1977) later reported a rapid digestion involving wet oxidation of fish tissue in HNO_3 with vanadium as catalyst, followed by AAS determination of mercury. Eleven collaborators compared it with the previous method on tuna fish samples of known mercury content and on spiked samples, and found it as accurate and precise, yet more rapid.

Burke and Albright (1971) applied AAS to the determination of copper and nickel in all types of tea. Seven collaborators analyzed six samples. Hoover (1972) reported a study in which eight laboratories determined lead in 24 samples of flour and cornmeal, apple and prune juice, hamburger and calf liver, and turnips and carrots. The lead was co-precipitated with strontium sulfate, converted to the carbonate, dissolved in nitric acid, and determined by AAS. AAS was also used to determine tin in foods (Elkins and Sulek 1979). Green beans, applesauce, and a fruit juice were fortified, and unfortified samples were used as controls. Samples were digested with HNO_3–H_2SO_4 mixture, methanol was added to enhance the signal, and the digestates were aspirated directly with nitrous oxide and the acetylene flame. Six collaborators obtained resulted acceptable at 5 ppm and above.

Osheim (1983) directed a collaborative study in which 12 laboratories determined copper in serum by AAS. Before the study began, methods for copper in serum used by a selected group of laboratories were surveyed. The laboratories were supplied with a Youden pair of bovine serum samples and requested to use their current method to determine serum copper. An AAS method was chosen on the basis of these results. The 12 collaborating laboratories each analyzed one blind duplicate and two Youden pairs of serum. A commercially available external control serum with a certified level of copper and a standard containing copper at 1000 mg/L were also furnished. The method required that the serum be diluted with equal quantities of distilled water and the standards with 10% glycerin to approximate the viscosity of the diluted serum.

Gajan and co-workers developed a group of methods in which the analyte could be determined by either AAS or polarography. A method for lead in fish (Gajan and Larry 1972) was tested by 19 laboratories, who found that the two determinative steps gave approximately the same results and could be used to confirm each other. Gajan et al. (1973) used a similar approach to determine cadmium in food. The method involved acid digestion, dithizone extraction, and determination by either AAS or polarography. Collaborators tested samples of lettuce, potatoes, orange juice, shredded wheat, milk, sugar, eggs, fish, rice, frankfurters, beans, and oysters spiked with cadmium at eight levels. Most of the collaborators used AAS as the determinative step; only three submitted polarographic data. The AAS method was adopted.

This dual approach was then used for the combination of AAS and anodic stripping voltammetry (ASV). Fiorino et al. (1973) directed a collaborative study in which 11 laboratories determined lead in evaporated milk. Five laboratories used both determinative steps, five used only AAS, and one used only ASV. Each collaborator analyzed two samples at each of five different levels plus two practice samples. Both techniques proved equally suitable and both were adopted by AOAC.

Holak (1980) conducted a collaborative study in which nine laboratories determined cadmium, lead, copper, arsenic, selenium, and zinc in strained chicken and apple sauce. Food samples were digested with nitric acid under pressure and aliquots of the solutions were analyzed by ASV after heating with equimolar $NaNO_3/KNO_3$ (cadmium, lead, copper) or by AAS after generation of hydrides (arsenic, selenium) or conventional AAS (zinc). The method was adopted for all metals except copper, for which there were too many outliers.

ASV alone was used as the determinative step in a method for lead and calcium in foods developed by Gajan et al. (1982). Food samples were dry ashed at 500 $\pm$ 50°C and the ash was dissolved in 2% nitric acid. Lead and cadmium were determined by ASV in a mixture of the sample solution and an acetate electrolyte at pH 4.3 $\pm$ 0.3. The collaborative study was reported by Capar et al. (1982). Twenty laboratories applied the method to strained green beans, beef (baby food), fish (mackerel), milk base infant formula, apple juice, and wheat farina. Every collaborator analyzed three of the commodities, each fortified with two duplicate lead and cadmium levels, for a total of four samples of each commodity. Thus 10 laboratories analyzed each commodity. Average accuracies of the results as determined by comparison to reference values were 96% for cadmium and 97% for lead.

Zink et al. (1983) used ASV to determine lead in evaporated milk and apple juice with no prior sample digestion. The method consisted of adding 2 mL of evaporated milk or 0.3 ml of fruit juice to 2.9 mL of a dechelating reagent. This mixture was then directly determined by ASV. Thirteen laboratories received 24 samples, two each at five different levels of analyte, together with duplicate practice samples of labeled lead content at each of two levels for each sample type. All unknowns were coded with random numbers. Approximately two-thirds of the reporting laboratories had never before determined lead in either evaporated milk or fruit juice.

In the early 1970s, a problem arose when ceramic food containers decorated with highly colored glazes containing lead and cadmium were found to represent a hazard. The containers were chiefly from foreign countries. When certain types of foods, e.g., highly acidic fruit juices, were stored in these containers, the toxic metals were leached out into the food and caused serious illness in the consumers of the food. Krinitz and Franco (1973) conducted a collaborative study of an AAS method for lead and cadmium extracted from the glazed surfaces of ceramic ware. Thirteen collaborators analyzed seven samples of 24-hr leach solution and obtained average recoveries for both metals of 101%. The method was also adopted by ASTM. This work was followed by a collaborative study of the effect of light on the leaching process (Krinitz and Holak 1976), in which the adopted method was modified to specify exposure to normal laboratory light for 8 to 10 hr during leaching, since eight collaborating laboratories found that additional cadmium was released in the presence of light. Krinitz (1978) developed a rapid test for detection of cadmium and lead extracted from glazed ceramic dinnerware that could be used as a field screening test. The test, a modification of the classical dithizone procedure, required no instrumentation. Fifteen laboratories used the 24-hr official method to analyze 90 samples field-tested for lead. Of the 63 samples found negative in the field test, 48 contained no lead after 24-hr leaching; others contained less than 0.7 ppm. Of 60 field-tested cadmium samples, 45 were negative; 30 of these showed no cadmium by the 24-hr test and 15 showed no more than 0.09 ppm.

In 1983, Gould et al. reported a collaborative study of a hot-leach method published by the World Health Organization for lead and cadmium in ceramic and enameled ware in which 14 laboratories participated. An acetic acid solution was heated at the boil for 2 hr in six samples of specially glazed ceramic ware and six samples of special enameled ware. The acid was allowed to cool and stand in contact with the ware for a further 22 hr. At the end of the 2-hr heating period and again at the end of 24 hr, the leach solution was assayed for lead and cadmium by AAS. Collaborators' results showed that the period of standing at room temperature was not necessary. Accordingly, the adopted method specified analysis of the leach solution immediately after the 2-hr heating period.

Mercury in fish and shellfish was again the subject of collaborative study by Hight and Capar in 1983. Methyl mercury was converted to the chloride by addition of HCl; the methyl mercuric chloride was extracted and determined by EC GLC on 5% DEGS-PS treated with inorganic mercuric chloride solution. Eight collaborating laboratories determined methyl mercury at two levels in blind duplicate samples of tuna, swordfish, oysters, and shrimp tissue. Hight (1987) modified this method and conducted a collaborative study in which eight laboratories determined methyl mercury in blind duplicate homogenates at two levels in tuna and one level in swordfish and oysters, and in single homogenates of swordfish and oysters containing methyl mercury at a second level. In the modified method, methyl mercury was isolated from homogenized, acetone-washed tissue by adding HCl, and the extracted methyl mercuric chloride was determined by EC GC.

C. Mycotoxins

In the early 1960s, a pressing need arose for methods to determine the mycotoxins, toxic metabolites formed from naturally occurring molds. The first group of mycotoxins to be the objective of an intensive program was the aflatoxins, derived from the mold *Aspergillus flavus*. The early methods involved their occurrence in peanuts and peanut products, and thus the methodology was studied under the topic "Nuts and Nut Products." Later mycotoxins became a separate topic included under "Natural Poisons," which occupied Chapter 26 of "Official Methods of Analysis." A symposium on various aspects of aflatoxins was held at the 1963 AOAC meeting.

In most of the early aflatoxin methods, the analytes were separated from the matrix by extraction, e.g., with aqueous methanol in a blender; the aflatoxins were washed out with $CHCl_3$, which was removed, and the residue was dissolved in an appropriate solvent. This solution was spotted on a thin layer chromatographic plate, and the four chief aflatoxins (B_1, B_2, G_1, and G_2) which separated were identified visually, usually by fluorescence under ultraviolet light. A 1-kg sample was required, and several techniques were applied to achieve a reduction in this sample requirement.

The first successful collaborative study was reported by Campbell and Funkhouser (1966). It was actually the second such study sponsored by the Aflatoxin Methodology Working Group of IUPAC; the first study resulted in considerable variation for peanuts and peanut meal. The second study was limited to peanut butter, and the method had been updated and modified. Thirteen collaborators from industry, government, and independent laboratories analyzed 12 samples fortified at 10 and 110 µg of aflatoxin B_1/kg; two naturally contaminated samples were also analyzed.

Stoloff (1967) conducted a collaborative study of a modification of the method of Andrellos and Reid (1964) for aflatoxin B_1 identification. The method was based on the altered chromatographic behavior of the aflatoxin after reaction with trifluoroacetic acid, formic acid/thionyl chloride, and glacial acetic acid/thionyl chloride, modified by an improved silica gel column cleanup and a clearer definition of sources of difficulty. Nineteen collaborators each examined three extracted samples, two of them naturally contaminated with aflatoxin B_1 and one aflatoxin-free extract to which an aflatoxin B_2 artifact had been added. No false identifications were made. Sixteen laboratories obtained reasonably good results with the trifluoroacetic acid and formic acid/thionyl chloride reagents; 12 laboratories obtained reasonably good results with the acetic acid/thionyl chloride reagent, but there was general difficulty with a side reaction assumed to be caused by inability to maintain anhydrous conditions.

Still working with peanut products, Eppley et al. (1968) conducted a collaborative study of the method for aflatoxin known as the CB method (for the Food and Drug Administration's Contaminants Branch). Thirteen laboratories analyzed naturally contaminated peanut butter and peanut meal, and peanut butter to which known amounts of aflatoxins B_1 and G_1 had been added. The study included preparatory isolation of the aflatoxin B_1 found in the naturally contaminated samples and confirmation of identity by chemical and biological tests. The method gave precision and accuracy equal to those of the earlier official method and was faster and more convenient. Pons (1969) conducted a 12-laboratory collaborative study of an extension of the basic procedure to cottonseed products in which fluorodensitometric measurement was an option. Levi (1969) directed a collaborative study in which nine laboratories detected aflatoxin B_1 in green coffee beans semiquantitatively.

Pohland et al. (1970) developed a rapid chemical confirmatory method for aflatoxin B_1, and Stoloff (1970) conducted a collaborative study by nine laboratories in which the earlier official method was compared to the revised version. In the revised method, aflatoxin-containing extracts were treated with concentrated HCl and water to yield a water adduct and with concentrated HCl and acetic anhydride to yield epimeric acetates. The collaborators analyzed 1-kg samples of a naturally contaminated peanut butter, a naturally contaminated corn meal, and a contaminated peanut meal. The revised method was shorter and simpler to perform and gave results as good as those by the official method or better.

Because of the difficulty of obtaining suitable aflatoxin standards for comparison with samples, Rodricks and Stoloff (1970) developed a procedure in which the concentration of solutions of aflatoxin standards was determined by their ultraviolet absorbance, and their purity was determined by examination of developed 50-ng spots on thin layer plates. Three collaborative studies were performed: one under IUPAC auspices and the others by Waltking (1970) and Levi (1969) and part of their own collaborative studies. Rodricks et al. (1970) also developed a method for criteria of purity of these standards based on their molar absorptivity values; it was adopted as part of the aflatoxin methodology.

Waltking (1970) proposed adoption of the "BF method" (for Best Foods Co.) widely accepted in industry for routine use in quality control of finished peanut butter. Fifteen collaborating laboratories compared it with the two official methods for naturally contaminated ground peanuts and specially prepared peanut butters as well as spiked commercial peanut butter and peanut meal. In the proposed method, samples were blended with aqueous methanol and extracted with $CHCl_3$. The solvent was evaporated and the extract was dissolved in benzene-acetonitrile and spotted on thin-layer plates. The collaborators found that the method gave as good results as the official methods, and was thus adopted as an alternative.

Extensions of aflatoxin methods to other commodities included collaborative studies of methods for copra, copra meal, and coconut by 19 laboratories (Baur and Armstrong 1971); cocoa beans by 10 laboratories (Scott and Przybylski 1971); and corn and soybeans by 15 laboratories, in which nine of the collaborators also calibrated standards and used densitometers in addition to visual evaluation (Shotwell and Stubblefield 1972). In 1973 Shotwell and Stubblefield conducted a collaborative study of three screening methods for aflatoxin in corn. All involved chromatography of partially purified extracts on minicolumns. Two of the methods were adopted: the method of Shannon et al. (1973) in which aflatoxin was extracted with acetone–water and impurities were precipitated with ammonium sulfate, and the method of Velasco (1972), which also used acetone-water as extractant but ferric gel as precipitant. DiProssimo (1974) compared two methods for aflatoxins in pistachio nuts in a collaborative study in which 13 laboratories used AOAC official methods I and III for peanut butter, with slight modifications, to analyze naturally contaminated, aflatoxin-free, and composites of pistachio nuts spiked with aflatoxins. Both methods were adopted; method III was simpler and faster but method I gave a cleaner extract. In 1975 Pons conducted a successful collaborative study in which 11 laboratories used a rapid modification of the official method for aflatoxins in cottonseed products to analyze six contaminated cottonseed meals, one blank meal, four contaminated kernel samples, and two ammonia-inactivated meals. Only one laboratory reported a false positive for the blank meal. The method was much faster than the official method.

A screening method for aflatoxins in mixed feeds, grains, nuts, and fruit products was studied by Romer (1975). The aflatoxins were extracted with acetone–water and interferences were removed by adding cupric carbonate and ferric chloride gel. The aflatoxins were extracted from the aqueous phase, and the washed extract was purified through a Velasco-type minicolumn, which captured the aflatoxins in a tight band. Thin-layer chromatography was used for quantitation of the extracts. Thirty-two collaborators from 10 countries tested 11 agricultural and food commodities (Romer and Campbell 1976). The method was adopted by IUPAC, American Association of Cereal Chemists, and AOAC.

A thin-layer chromatographic method for the determination of aflatoxin B_1 in eggs was developed by Trucksess et al. (1977) and collaboratively studied by Nesheim and Trucksess (1978). Samples were extracted with acetone and lipids removed with lead acetate solution. Samples were reextracted with petroleum ether, transferred to $CHCl_3$, and eluted from a silica gel column with $CHCl_3$. Two-dimensional thin-layer chromatography was used for the determinative step, with either visual or densitometric quantitation. Thirteen collaborators analyzed three known practice samples and nine unknowns containing added aflatoxin B_1 at four levels.

Minicolumn detection methods were applied to almonds in a collaborative study reported by Stanley et al. (1979). Twenty laboratories took part in Phase I of the study in which five almond butter samples and two practice samples stated to contain 0 and 10 ng aflatoxin/g were analyzed. Phase II, which was not performed by all laboratories, consisted of two more coded samples duplicating the original blank and either a 5 or a 10 ng/g sample. Collaborators correctly identified 96% of the samples containing 5 to 25 ng/g total aflatoxins and 83% of negative samples. Shannon and Shotwell (1979) also applied the minicolumn technique to aflatoxin in yellow corn. They conducted a comparison of the CPC modified method, the Holaday modified method, and a combination of the two. Nineteen laboratories participated; 89% detected 10 ng/g by the Holaday-Velasco combined method in naturally contaminated yellow corn at several levels. Pons et al. (1980) revised the method for aflatoxins in cottonseed products by reducing the quantities of material in the column chromatography and changing the eluting solvents, and by permitting use of liquid chromatography as an alternative to thin-layer chromatography in the determinative step. The 12 participating laboratories found no significant difference in the results by either procedure, although liquid chromatography did reduce the between-laboratory error.

In a return to the problem of aflatoxins in peanut products, Stack (1974) undertook a collaborative comparison of AOAC methods I and II for aflatoxins in peanut butter. Seven collaborators tested the methods on two naturally contaminated peanut butters and eight spiked samples. Both methods were acceptable but method I gave greater accuracy and precision and was subject to fewer interferences. Shotwell and Holaday (1981) continued the study of minicolumn detection methods. Fifteen collaborators compared the Holaday–Velasco method and a modified Holaday method on raw peanuts. The Holaday–Velasco method gave fewer false negatives and false positives.

Confirmation of identity of aflatoxins was the subject of two collaborative studies. Stack and Pohland (1975) used a chemical method involving the formation of a derivative directly on the thin-layer plate, which was based on the catalytic action of trifluoroacetic acid upon addition of water across the double bond of the terminal furan ring of aflatoxin B_1, converting it to aflatoxin B_{2a}. Aflatoxin G_1 was similarly converted to G_{2a}. Park et al. (1985) used negative ion chemical ionization mass spectrometry to confirm identity of aflatoxin B_1.

Laboratories in the United States, England, and West Germany examined 12 partially purified, dry film extracts from naturally and artificially contaminated roasted peanuts, cottonseed, and ginger root containing varying quantities of the aflatoxin. The extracts required additional cleanup before mass spectrometry. Identity was confirmed in 19.5, 90.9, and 100% of samples containing less than 5, 5 to 10, and more than 10 ng/g of aflatoxin B_1, respectively, by solid probe introduction using full mass scans.

Another aflatoxin, M_1, was also the subject of several international collaborative studies. In the first study (Stubblefield and Shannon 1974) 19 laboratories assayed liquid and powdered milk, cheese, and butter containing aflatoxin M_1 by the quantitative method of Pons et al. (1973), with modifications by Stubblefield and Shannon, together with the chemical confirmation of Stack et al. (1975). In 1980 Stubblefield et al. reported a second study by 23 collaborators to test an improved version of the method for the rapid determination and thin-layer confirmation of aflatoxin M_1 identity in dairy products. Another such study (Stubblefield and Kwolek 1986) involved 14 laboratories in 5 countries who successfully used the rapid liquid chromatography method of Ferguson-Foos and Warren (1984) in both normal and reversed phase. The reversed phase method was adopted. In 1982 Stubblefield et al. directed a collaborative study of aflatoxins B_1 and M_1 in artificially contaminated beef liver in which 13 laboratories tested two thin-layer chromatography methods for confirmation of identity; both were adopted. Two determinative methods using two-dimensional thin-layer chromatography were also tested; one required an internal standard and the other was a direct comparison with standards on thin-layer plates. The internal standard was found to give no better results, and therefore the direct method was adopted.

Methods for other mycotoxins besides the aflatoxins have been collaboratively studied and adopted. Stack and Rodricks (1973) conducted a study of a quantitative method for sterigmatocystin in grains in which 17 laboratories examined corn, wheat, and barley. Samples were extracted with acetonitrile–water and partitioned first into hexane and then into $CHCl_3$. Further cleanup, if needed, was provided by silica gel column chromatography. Sterigmatocystin was estimated by comparing its fluorescence on a thin layer plate with that of a standard. The method was adopted for barley and wheat. Nesheim et al. (1973) described a method for ochratoxins A and B and their esters in barley. Thirteen collaborators each analyzed one blank, two spiked samples, and three naturally contaminated samples. The samples were extracted with water–$CHCl_3$ and the extracts were chromatographed on aqueous $NaHCO_3$–diatomaceous earth columns. The esters were removed with hexane–$CHCl_3$ and the ochratoxins were eluted with formic acid-$CHCl_3$ and purified further on columns of aqueous $NaHCO_3$–methanol-diatomaceous earth. The extracts were then quantitated by the intensity of their fluorescence on thin-layer plates. The method was adopted as quantitative for ochratoxin A and qualitative for the other compounds. Levi (1975) modified the method and applied it to green coffee. Eleven laboratories analyzed seven samples spiked at three levels.

In a collaborative study conducted by Scott (1974), patulin was determined by 20 laboratories in 9 spiked samples of apple juice, two of them blanks. The juice was extracted with ethyl acetate and the extract was cleaned up on a silica gel column. The eluate was concentrated and spotted on thin-layer plates. Patulin was determined visually after the plates were sprayed with 3-methyl-2-benzothiazolinone hydrazone·HCl solution. The method was adopted as a semiquantitative determination.

Shotwell et al. (1976) conducted a collaborative study of the method of Eppley (1968) for zearalenone in which 16 laboratories reported results on corn spiked at four levels. More recently Bennett et al. (1985) reported a collaborative study in which 13 laboratories analyzed seven corn samples (two blanks and five samples fortified with the two analytes) for α-zearalenol and zearalenone. All collaborators detected both mycotoxins at 50 ng/g; three collaborators reported false positives for both.

Two methods for deoxynivalenol in wheat have been studied collaboratively. Eppley et al. (1986) described a thin-layer chromatographic method in which the sample was extracted with acetonitrile–water and cleaned up on a disposable column of charcoal, Celite, and alumina; the extracts were examined on thin layer plates sprayed with aluminum chloride solution. Fifteen collaborators analyzed 12 samples, 2 naturally contaminated and 10 spiked. Ware et al. (1986) reported a gas chromatographic determination with EC detection. Ten laboratories analyzed six samples in duplicate: three spiked at different levels, a control, and two naturally contaminated samples. Recovery from the spiked samples averaged 92.2%.

Stoloff (1980) published an excellent review of the history of the mycotoxin problem and subsequent regulatory efforts.

Summary

The Association of Official Analytical Chemists, founded more than a century ago, is still carrying out its mission of providing reliable, validated methods of analysis for materials important to agriculture and the public health. Although the Association has greatly expanded the scope and number of its activities and its structure and governance have become more complex, it still relies on the collaborative study as the best approach to providing these methods. Studies in three areas—pesticide residues, metals in foods, and mycotoxins—amply illustrate this important concept. During the past quarter of a century, the period emphasized in this review, methods for pesticide residues changed from determination of single pesticides by chemical reactions and spectrophotometry to multiresidue analysis in which a number of pesticides are determined simultaneously by a combination of extraction, cleanup, and quantitation, chiefly by gas chromatography with a variety of detectors. Methods for metals in foods still tend to concentrate on the traditional metallic contaminants of arsenic, cadmium, mercury, and lead, although methods for additional metals have been studied collaboratively and adopted. Various techniques have been used but in recent years atomic absorption spectrophotometry and anodic stripping voltammetry have predominated. The mycotoxin methods rely very little on instrumental detection. Rather, these naturally formed mold metabolites are usually detected by their fluorescence on thin-layer chromatographic plates. Variations in methodology chiefly concern extraction and cleanup techniques. Because the need for reliable methods seems to continue unabated, the work of the Association of Official Analytical Chemists is likely to fulfill a need well into its second century of existence.

References

Andrellos PJ, Reid GR (1964) Confirmatory tests for aflatoxin B_1. J Assoc Offic Agric Chem 47:801–803.

Ault JA, Spurgeon TE (1984) Multiresidue gas chromatographic method for determinating organochlorine pesticides in poultry fat: Collaborative study. J Assoc Offic Anal Chem 67:284–289.

Bartlet JC (1964) Summary of results of the collaborative tests on the Rhodamine B colorimetric method for antimony. J Assoc Offic Agric Chem 47:630–632.

Baur FJ, Armstrong JC (1971) Collaborative study of a modified method for the determination of aflatoxins in copra, copra meal, and coconut. J Assoc Offic Anal Chem 54:874–878.

Bennett GA, Shotwell OL, Kwolek WF (1985) Liquid chromatographic determination of *alpha*-zearalenol and zearalenone in corn: Collaborative study. J Assoc Offic Anal Chem 68:958–961.

Bluman N (1964) Phosdrin residues in fruits and vegetables. J Assoc Offic Agric Chem 47:272–280.

Bong RL (1977) Collaborative study of the recovery of hexachlorobenzene and mirex in butterfat and fish. J Assoc Offic Anal Chem 60:229–232.

Bowman MC (1984) Pesticides, their analysis and the AOAC: An overview of a century of progress (1884–1984). J Assoc Offic Anal Chem 67:204–209.

Burke JA (1968) Report of the General Referee on Chlorinated Pesticides. J Assoc Offic Anal Chem 51:311–314.

Burke JA (1980) Report of the General Referee on Organochlorine Pesticides. J Assoc Offic Anal Chem 63:277–282.

Burke KE, Albright CH (1971) Atomic absorption spectrophotometry for determination of parts per million quantities of copper and nickel in tea. J Assoc Offic Anal Chem 54:658–662.

Butrill WH (1973) Collaborative study of a colorimetric method for determining arsenic residues in red meat and poultry. J Assoc Offic Anal Chem 56:1144–1148.

Campbell AD, Funkhouser JT (1966) Collaborative study on analysis of aflatoxins in peanut butter. J Assoc Offic Anal Chem 49:730–739.

Capar SG, Gajan RJ, Madszar E, Albert RH, Sanders M, Zyren J (1982) Determination of lead and cadmium in foods by anodic stripping voltammetry. II. Collaborative study. J Assoc Offic Anal Chem 65:978–986.

Carr RL (1970) Collaborative study of the Mills method for multiple chlorinated pesticide residues in butterfat. J Assoc Offic Anal Chem 53:152–154.

Carr RL (1971) Collaborative study of a method for multiple chlorinated pesticide residues in fish. J Assoc Offic Anal Chem 54:525–527.

Crist HL, Moseman RF, Noneman JW (1975) Rapid determination and confirmation of low levels of hexachlorobenzene in adipose tissue. Bull Environ Contam Toxicol 14:273–279.

Davidson AW (1966) Collaborative study of a rapid method for multiple chlorinated pesticide residues in small fruits. J Assoc Offic Anal Chem 49:468–472.

DiProssimo (1974) Collaborative study comparing two methods for the determination of aflatoxins in pistachio nuts. J Assoc Offic Anal Chem 57:1114–1120.

Elkins ER, Sulek A (1979) Atomic absorption determination of tin in foods. J Assoc Offic Anal Chem 62:1050–1053.

Eppley RM (1968) Screening method for zearalenone, aflatoxin, and ochratoxin. J Assoc Offic Anal Chem 51:74–78.

Eppley RM, Stoloff L, Campbell AD (1968) Collaborative study of "A versatile procedure for assay of aflatoxins in peanut products," including preparatory separation and confirmation of identity. J Assoc Offic Anal Chem 51:67–73.

Eppley RM, Trucksess MW, Nesheim S, Thorpe CW, Pohland AE (1986) Thin layer chromatographic method for determination of deoxynivalenol in wheat: Collaborative study. J Assoc Offic Anal Chem 69:37–40.

Erney R (1983) Rapid screening procedure for pesticides and polychlorinated biphenyls in fish: Collaborative study. J Assoc Offic Anal Chem 66:969–973.

Ferguson-Foos J, Warren JD (1984) Improved cleanup for liquid chromatographic analysis and fluorescent detection of aflatoxins M_1 and M_2 in fluid milk products. J Assoc Offic Anal Chem 67:1111–1114.

Finsterwalder CE (1976) Collaborative study of an extension of the Mills et al. method for the determination of pesticide residues in foods. J Assoc Offic Anal Chem 59:169–171.

Fiorino JA, Moffitt RA, Woodson AL, Gajan RJ, Huskey GE, Scholz RG (1973) Determination of lead in evaporated milk by atomic absorption spectrophotometry and anodic stripping voltammetry: Collaborative study. J Assoc Offic Anal Chem 56:1246–1251.

Gajan RJ (1969) Collaborative study of confirmative procedures by single sweep oscillographic polarography for the determination of organophosphorus pesticide residues in nonfatty foods. J Assoc Offic Anal Chem 52:811–817.

Gajan RJ, Larry D (1972) Determination of lead in fish by atomic absorption spectrophotometry and by polarography. I. Development of the methods. II. Collaborative study. J Assoc Offic Anal Chem 55:727–732, 733–736.

Gajan RJ, Capar SG, Subjoc CA, Sanders M (1982) Determination of lead and cadmium in foods by anodic stripping voltammetry. I. Development of method. J Assoc Offic Anal Chem 65:970–977.

Gajan RJ, Gould JF, Watts JO, Fiorino JA (1973) Collaborative study of a method for the atomic absorption spectrophotometric and polarographic determination of cadmium in food. J Assoc Offic Anal Chem 56:876–881.

Gaul J (1966) Collaborative study of a method for multiple chlorinated pesticides residues in leafy and cole-type vegetables. J Assoc Offic Anal Chem 49:463–467.

Guiffrida L, Ives NF, Bostwick DC (1966) Gas chromatography of pesticides: Improvements in the use of special ionization detection systems. J Assoc Offic Anal Chem 49:8–21.

Goodwin ES, Goulden R, Reynolds JG (1961) Rapid identification and determination of residues of chlorinated pesticides in crops by gas-liquid chromatography. Analyst 86:697–709.

Gould JH, Butler SW, Boyer KW, Steele EA (1983) Hot leaching of ceramic and enameled cookware: Collaborative study. J Assoc Offic Anal Chem 66:610–619.

Guidelines for Collaborative Study Procedure to Validate Characteristics of a Method of Analysis (1988) IUPAC Workshop on Harmonization of Collaborative Analytical Studies, Geneva, Switzerland, May 4–5, 1987; J Assoc Offic Anal Chem 71:161–172.

Helrich K (1984) The great collaboration. Association of Official Analytical Chemists, Arlington, VA.

Hight SC (1987) Rapid determination of methyl mercury in fish and shellfish: Collaborative study. J Assoc Offic Anal Chem 70:667–672.

Hight SC, Capar SG (1983) Electron capture gas-liquid chromatographic determination of methyl mercury in fish and shellfish: Collaborative study. J Assoc Offic Anal Chem 66:1121–1128.

Holak W (1980) Analysis of foods for lead, cadmium, copper, zinc, arsenic, and selenium, using closed system sample digestion: Collaborative study. J Assoc Offic Anal Chem 63:485–495.

Holden ER (1975) Collaborative study of the 2,4-dinitrophenyl ether multiresidue method for use in determining four carbamate pesticides in crops. J Assoc Offic Anal Chem 58:562–565.

Hoover WL (1972) Collaborative study of a method for determining lead in plant and animal products. J Assoc Offic Anal Chem 55:737–740.

Horwitz W (1964) The Association of Official Agricultural Chemists (AOAC). Residue Reviews, Vol. 7:37–60.

Ihnat M (1974) Collaborative study of a fluorimetric method for determining selenium in foods. J Assoc Offic Anal Chem 57:373–378.

Johnson DP (1964) Determination of Sevin insecticide residues in fruits and vegetables. J Assoc Offic Anal Chem 47:280–286.

Johnson LY (1962) Separation of dieldrin and endrin from other chlorinated pesticide residues. J Assoc Offic Agric Chem 45:363–365.

Johnson LY (1965) Collaborative study of a method for multiple chlorinated pesticide residues in fatty foods. J Assoc Offic Agric Chem 48:668–675.

Kovacs MF Jr. (1963) Thin layer chromatography for chlorinated pesticide residue analysis. J Assoc Offic Agric Chem 46:884–893.

Krause RT (1966) Collaborative study of a method for chlorinated pesticide residues in nonfatty vegetables. J Assoc Offic Anal Chem 49:460–463.

Krause RT (1973) Determination of several chlorinated pesticides by the AOAC multiresidue method with additional quantitation of Perthane after dehydrochlorination: Collaborative study. J Assoc Offic Anal Chem 56:721–727.

Krause RT (1985) Liquid chromatographic determination of N-methylcarbamate insecticides and metabolites in crops. I. Collaborative study. J Assoc Offic Anal Chem 68:726–733.

Krinitz B, Franco V (1973) Collaborative study of an atomic absorption method for the determination of lead and cadmium extracted from glazed ceramic surfaces. J Assoc Offic Anal Chem 56:864–875.

Krinitz B, Holak W (1974) Simple, rapid digestion technique for the determination of mercury in seafood by flameless atomic absorption spectrophotometry. J Assoc Offic Anal Chem 57:568–569.

Krinitz B, Holak W (1976) Collaborative study of effect of light on cadmium and lead leaching from ceramic glazes. J Assoc Offic Anal Chem 59:158–161.

Krinitz B (1978) Rapid screening field test for detecting cadmium and lead extracted from glazed ceramic dinnerware: Collaborative study. J Assoc Offic Anal Chem 61:1124–1129.

Laski RR (1974) Collaborative study of a method for organophosphorus pesticide residues in apples and green beans. J Assoc Offic Anal Chem 57:930–933.

Levi C (1969) Collaborative study on a method for detection of aflatoxin B_1 in green coffee beans. J Assoc Offic Anal Chem 52:1300–1303.

Levi CP (1975) Collaborative study of a method for the determination of ochratoxin A in green coffee. J Assoc Offic Anal Chem 58:258–262.

Lovelock JE, Lipsky SR (1960) Electron affinity spectroscopy: A new method for the identification of functional groups in chemical compounds separated by gas chromatography. J Am Chem Soc 82:431.

Luke MA, Froberg JE, Doose GM, Masumoto HT (1981) Improved multiresidue gas chromatographic determination of organophosphorus, organonitrogen, and organohalogen pesticides in produce, using flame photometric and electrolytic conductivity detectors. J Assoc Offic Anal Chem 64:1187–1195.

Martin R, Schwartzman G (1964) Determination of nicotine residues on foods. J Assoc Offic Agric Chem 47:303–305.

Mills PA (1961) Collaborative study of certain chlorinated organic pesticides in dairy products. J Assoc Offic Agric Chem 44:171–177.

Mitchell LR (1976) Collaborative study of the determination of endosulfan, endosulfan sulfate, tetrasul, and tetradifon residues in fresh fruits and vegetables. J Assoc Offic Anal Chem 59:202–212.

Munns RK, Holland DC (1971) Determination of mercury in fish by flameless atomic absorption: Collaborative study. J Assoc Offic Anal Chem 54:202–205.

Munns RK, Holland DC (1977) Rapid digestion and flameless atomic absorption spectrophotometry of mercury in fish: Collaborative study. J Assoc Offic Anal Chem 60:833–837.

Nesheim S, Hardin NF, Francis OJ Jr, Langham WS (1973) Analysis of ochratoxins A and B and their esters in barley. I. Development of the method. II. Collaborative study. J Assoc Offic Anal Chem 56:817–821.

Nesheim S, Trucksess MW (1978) Thin Layer chomatographic determination of aflatoxin B_1 in eggs: Collaborative study. J Assoc Offic Anal Chem 61:569–573.

Onley JH (1977) Gas-liquid chromatographic method for determining ethylene thiourea in potatoes, spinach, applesauce, and milk: Collaborative study. J Assoc Offic Anal Chem 60:1111–1115.

Osheim DL (1983) Atomic absorption determination of serum copper: Collaborative study. J Assoc Offic Anal Chem 66:1140–1142.

Palmer NJ, Benson WR (1968) Collaborative study of the thin layer chromatographic method for carbaryl residues in apples and spinach. J Assoc Offic Anal Chem 51:679–681.

Park DL, DiProssimo V, Abdel-Malek E, Trucksess MW, Nesheim S, Brumley WC, Sphon JA, Barry TL, Petzinger G (1985) Negative ion chemical ionization mass spectrometric method for confirmation of identity of aflatoxin B_1: Collaborative study. J Assoc Offic Anal Chem 68:636–640.

Pasarela NR (1964) Total dodine residues in fruits. J Assoc Offic Agric Chem 47:300–303.

Pohland AE, Yin L, Dantzman JG (1970) Rapid chemical confirmatory method for aflatoxin B_1. I. Development of the method. J Assoc Offic Anal Chem 53:101–102.

Pons WA Jr (1969) Aflatoxins in cottonseed products: Collaborative study. J Assoc Offic Anal Chem 52:61–72.

Pons WA Jr (1975) Collaborative study of a rapid method for determining aflatoxins in cottonseed products. J Assoc Offic Anal Chem 58:746–753.

Pons WA Jr, Cucullu AF, Franz AO Jr, Lee LS, Goldblatt LA (1973) Rapid detection of aflatoxin contamination in agricultural products. J Assoc Offic Anal Chem 56:803–807.

Pons WA Jr, Lee LS, Stoloff L (1980) Revised method for aflatoxins in cottonseed products, and comparison of thin layer and high performance liquid chromatography determinative steps: Collaborative study. J Assoc Offic Anal Chem 63:899–906.

Randall RC (1970) Ultraviolet determination of naphthalene acetic acid in apples and potatoes. J Assoc Offic Anal Chem 53:149–151.

Rodricks JV, Stoloff L (1970) Determination of concentration and purity of aflatoxin standards. J Assoc Offic Anal Chem 53:92–95.

Rodricks JV, Stoloff L, Pons WA Jr, Robertson JA, Goldblatt LA (1970) Molar absorptivity values for aflatoxins and justification for their use as criteria of purity of analytical standards. J Assoc Offic Anal Chem 53:96–101.

Rogers GR (1968) Collaborative study of atomic absorption spectrophotometric method for determining zinc in foods. J Assoc Offic Anal Chem 51:1042–1045.

Romer TR (1975) Screening method for the detection of aflatoxins in mixed feeds and other agricultural commodities with subsequent confirmation and quantitative measurement of aflatoxins in postive samples. J Assoc Offic Anal Chem 58:500–506.

Romer TR, Campbell AD (1976) Collaborative study of a screening method for the detection of aflatoxins in mixed feeds, other agricultural products, and foods. J Assoc Offic Anal Chem 59:110–117.

Sawyer LD (1985) The Luke et al. method for determining multipesticide residues in fruits and vegetables: Collaborative study. J Assoc Offic Anal Chem 68:64–71.

Sawyer LD, Walters SM (1986) Gas chromatographic method for ethylene dibromide in grains and grain-based products: Collaborative study. J Assoc Offic Anal Chem 69:847–851.

Scott PM (1974) Collaborative study of a chromatographic method for determination of patulin in apple juice. J Assoc Offic Anal Chem 57:621–626.

Scott PM, Przybylski W (1971) Collaborative study of a method for the analysis of cocoa beans for aflatoxins. J Assoc Offic Anal Chem 54:540–544.

Shannon GM, Shotwell OL (1979) Minicolumn detection methods for aflatoxin in yellow corn: Collaborative study. J Assoc Offic Anal Chem 62:1070–1075.

Shannon GM, Stubblefield RD, Shotwell OL (1973) Modified rapid screening method for aflatoxin in corn. J Assoc Offic Anal Chem 56:1024–1025.

Shotwell OL, Goulden ML, Bennett GA (1976) Determination of zearalenone in corn: Collaborative study. J Assoc Offic Anal Chem 59:666–670.

Shotwell OL, Holaday CE (1981) Minicolumn detection methods for aflatoxin in raw peanuts: Collaborative study. J Assoc Offic Anal Chem 64:674–677.

Shotwell OL, Stubblefield RD (1972) Collaborative study of the determination of aflatoxin in corn and soybeans. J Assoc Offic Anal Chem 55:781–788.

Shotwell OL, Stubblefield RD (1973) Collaborative study of three screening methods for aflatoxin in corn. J Assoc Offic Anal Chem 56:808–812.

Stack ME (1974) Collaborative study of AOAC methods I and III for the determination of aflatoxins in peanut butter. J Assoc Offic Anal Chem 57:871–874.

Stack ME, Pohland AE (1975) Collaborative study of a method for chemical confirmation of the identity of aflatoxin. J Assoc Offic Anal Chem 58:110–113.

Stack ME, Rodricks JV (1973) Collaborative study of quantitative determination and chemical confirmation of sterigmatocystin in grains. J Assoc Offic Anal Chem 56: 1123–1125.

Stanley GI Jr, DiProssimo VP, Koontz AC (1979) Minicolumn detection methods applied to almonds: Collaborative study. J Assoc Offic Anal Chem 62:136–140.

Stoloff L (1967) Collaborative study of a method for the identification of aflatoxin B_1 by derivative formation. J Assoc Offic Anal Chem 50:354–360.

Stoloff L (1970) Rapid chemical confirmatory method for aflatoxin B_1. II. Collaborative study. J Assoc Offic Anal Chem 53:102–104.

Stoloff L (1980) Aflatoxin control: Past and present. J Assoc Offic Anal Chem 63:1067–1073.

Storherr RW, Watts RR (1965) A sweep co-distillation cleanup method for organophosphate pesticides. J Assoc Offic Agric Chem 48:1154–1158.

Storherr RW, Watts RR (1968) Collaborative study of the ethyl acetate extraction, sweep co-distillation cleanup, and gas-liquid chromatographic determination, using six parent organophosphate pesticides. J Assoc Offic Anal Chem 51:662–665.

Storherr RW, Murray EJ, Klein I, Rosenberg LA (1967) Sweep co-distillation cleanup of fortified edible oils for determination of organophosphate and chlorinated hydrocarbon pesticides. J Assoc Offic Anal Chem 50:605–615.

Stubblefield RD, Kwolek (1986) Rapid liquid chromatographic determination of aflatoxins M_1 and M_2 in artificially contaminated fluid milks: Collaborative study. J Assoc Offic Anal Chem 69:880–885.

Stubblefield RD, Shannon GM (1974) Collaborative study of methods for the determination and chemical confirmation of aflatoxin M_1 in dairy products. J Assoc Offic Anal Chem 57:852–857.

Stubblefield RD, Kwolek WF, Stoloff L (1982) Determination and thin layer chromatographic confirmation of identity of aflatoxin B_1 and M_1 in artificially contaminated beef livers: Collaborative study. J Assoc Offic Anal Chem 65:1435–1444.

Stubblefield RD, Van Egmond HP, Paulsch WE, Schuller PL (1980) Determination and confirmation of identity of aflatoxin M_1 in dairy products: Collaborative study. J Assoc Offic Anal Chem 63:907–921.

Trucksess MW, Stoloff L, Pons WA Jr, Cucullu AF, Lee LS, Franz AO Jr (1977) J Assoc Offic Anal Chem 60:795–798.

Velasco J (1972) J Am Oil Chem Soc 49:141–142.

Waltking AE (1970) Collaborative study of three methods for determination of aflatoxin in peanuts and peanut products. J Assoc Offic Anal Chem 53:104–113.

Ware GM, Francis OJ, Carman AS, Kuan SS (1986) Gas chromatographic determination of deoxynivalenol in wheat with electron capture: Collaborative study. J Assoc Offic Anal Chem 69:899–901.

Watts JO, Klein AK (1962) Determination of chlorinated pesticide residues by electron capture gas chromatography. J Assoc Offic Anal Chem 45:102–108.

Watts RR, Storherr RW (1967) Sweep co-distillation cleanup of milk for determination of organophosphate and chlorinated hydrocarbon pesticides. J Assoc Offic Anal Chem 50:581–585.

Watts RR, Hodgson DW, Crist HL, Moseman RF (1980) Improved method for hexachlorobenzene and mirex determination with hexachlorobenzene confirmation in adipose tissue: Collaborative study. J Assoc Offic Anal Chem 63:1128–1134.

Wessel JR (1967) Collaborative study of a method for multiple organophosphorus pesticide residues in nonfatty foods. J Assoc Offic Anal Chem 50:430–439.

Woolson EA (1973) Extraction of chlorinated hydrocarbon insecticides from soil: Collaborative study. J Assoc Offic Anal Chem 56:728.

Youden WJ (1970) Statistical techniques for collaborative tests. Association of Official Analytical Chemists, Arlington, VA.

Youden WJ, Steiner EH (1975) Statistical manual of the AOAC. Association of Official Analytical Chemists, Arlington, VA.

Zink EW, Davis PH, Griffin RM, Matson WR, Moffitt RA, Sakai DT (1983) Direct determination of lead in evaporated milk and apple juice by anodic stripping voltammetry: Collaborative study. J Assoc Office Anal Chem 66:1414–1420.

Manuscript received September 18, 1988; accepted September 26, 1988.

Subject Index

Reviews of Environmental Contamination and Toxicology

Edited by

George W. Ware

Published by

Springer-Verlag New York • Berlin • Heidelberg • Tokyo

The original copy and one good photocopy of the manuscript, complete with figures and tables, are required. Manuscripts will be published in the order in which they are received, reviewed, and accepted. They should be sent to the editor:

> Dr. George W. Ware
> College of Agriculture
> University of Arizona
> Tucson, Arizona 85721
> Telephone: (602) 621-3859 (office)
> (602) 299-3735 (home)

1. Manuscript

The manuscript, in English, should be typewritten, double-spaced throughout (including reference section), on one side of 8½ × 11-inch blank white paper, with at least one-inch margins. The first page of the manuscript should start with the title of the manuscript, name(s) of author(s), with author affiliation(s) as first-page starred footnotes, and "Contents" section. Pages should be numbered consecutively in arabic numerals, including those bearing figures and tables only. In titles, in-text outline headings and subheadings, figure legends, and table headings only the initial word, proper names, and universally capitalized words should be capitalized.

Footnotes should be inserted in text and numbered consecutively in the text using arabic numerals.

Tables should be typed on separate sheets and numbered consecutively within the text in *roman numerals*; they should bear a descriptive heading, in lower case, which is underscored with one line and which starts after the word "Table" and the appropriate roman numeral; *footnotes in tables* should be designated consecutively within a table by the lower-case alphabet. *Figures* (including photos, graphs, and line drawings) should be numbered consecutively within the text in arabic numerals; each figure should be affixed to a separate page bearing a legend (below the figure) in lower case starting with the term "Fig." and a number.

2. Summary

A concise but informative summary (double-spaced) must conclude the text of each manuscript; it should summarize the significant content and major conclusions presented. It must not be longer than two 8½ × 11-inch pages of double-spaced typing. As a summary, it should be more informative than the usual abstract.

3. References

All papers, books, and other works cited in the text must be included in a "References" section (*also double-spaced*) at the end of the manuscript. If comprehensive papers on the same subject have been published, they should be cited when the bibliographic citations extend farther back than to these papers.

All papers cited in the text should be given in parentheses and alphabetically when more than one reference is cited at a time, e.g. (Coats and Smith 1979, Holcombe et al. 1982, Stratton 1986), except when the author is mentioned, as for example, "and the study of Roberts and Stoydin (1985)." References to unpublished works should be kept to a minimum and mentioned only in the text itself in parentheses. References to published works are given at the end of the text in alphabetical order under the first author's name, citing all authors (surnames followed by initials throughout; do not use "and") according to the following examples:

Periodicals: Name(s), initials, year of publication in parentheses, full article title, journal title as abbreviated in "The ACS Style Guide: A Manual for Authors and Editors" of the American Chemical Society, volume number, colon, first and last page numbers. Example:

Leistra MT (1970) Distribution of 1,3-dichloropropene over the phases in soil. J Agric Food Chem 18:1124–1126.

Books: Name(s), initials, year of publication in parentheses, full title, edition, volume number, name of publisher, place of publication, first and last page numbers. Example:

Gosselin R, Hodge H, Smith R, Gleason M (1976) Clinical Toxicology of Commercial Products, 4th Ed. Wilkins-Williams, Baltimore, MD, pp 119–121.

Work in an edited collection: Name(s), initials, year of publication in parentheses, full title. In: name(s) and initial(s) of editor(s), the abbreviation ed(s) in parentheses, name of publisher, place of publication, first and last page numbers. Example:

Metcalf RL (1978) Fumigants. In: White-Stevens J (ed) Pesticides in the environment. Marcel Dekker, New York. pp 120–130.

References by the same author(s) are arranged chronologically. If more than one reference by the same author(s) published in the same year is cited, use a, b, c after year of publication in both text and reference list.

4. Illustrations

Illustrations may be included only when indispensable for the comprehension of text. They should not be used in place of concise explanations in text. Schematic line drawings must be drawn carefully. For other illustrations, clearly defined black-and-white glossy photos are required. Should darts (arrows) or letters be required on a photo or other type of illustration, they should be marked neatly with a soft pencil on a duplicate copy or on an overlay, with the end of each dart indicated by a fine pinprick; darts and lettering will be transferred to the illustrations by the publisher.

Photos should not be less than five × seven inches in size. Alterations of photos in page proof stage are not permitted. *Each photograph or other illustration should be marked on the back, distinctly but lightly, with soft pencil, with first author's name, figure number, manuscript page number, and the side which is the top.*

If illustrations from published books or periodicals are used, the exact source of each should be included in the figure legend: if these "borrowed" illustrations are copyrighted by others, permission of the copyright holder to reproduce the illustration must be secured by the author.

5. Chemical Nomenclature

All pesticides and other subject-matter chemicals should be identified according to *Chemical Abstracts*, with the full chemical name in text in parentheses or brackets the first time a common or trade name is used. *If many such names are used, a table of the names, their precise chemical designations, and their* Chemical Abstract Numbers (CAS) *should be included as the last table in the manuscript, with a numbered footnote reference to this fact on the first text page of the manuscript.*

6. Miscellaneous

Abbreviations: Common units of measurement and other commonly abbreviated terms and designations should be abbreviated as listed below; if any others are used often in a manuscript, they should be written out the first time used, followed by the normal and acceptable abbreviation in parentheses [e.g., Acceptable Daily Intake (ADI), Angstrom (Å), picogram (pg)]. Except for inch (in.) and number (no., when followed by a numeral), abbreviations are used without periods. Temperatures should be reported as "°C" or "°F" (e.g., mp 41° to 43°C). Because the metric system is the international standard, when pounds (lb) and gallons (gal) are used the metric equivalent should follow in parentheses.

A	acre		sec	second(s)
bp	boiling point		μg	microgram(s)
cal	calorie		μL	microliter(s)
cm	centimeter(s)		μm	micrometer(s)
cu	cubic (as in "cu m")		mg	milligram(s)
d	day		mL	milliliter(s)
ft	foot (feet)		mm	millimeter(s)
gal	gallon(s)		m$\underline{M}$	millimolar
g	gram(s)		min	minute(s)
ha	hectare		$\underline{M}$	molar
hr	hour(s)		mon	month(s)
in.	inch(es)		ng	nanogram(s)
id	inside diameter		nm	nanometer(s) (millimicron)
kg	kilogram(s)		$\underline{N}$	normal
L	liter(s)		no.	number(s)
mp	melting point		od	outside diameter
m	meter(s)		oz	ounce(s)
ppb	parts per billion		sp gr	specific gravity
ppm	parts per million		sq	square (as in "sq m")
ppt	parts per trillion		vs	versus
pg	picogram		wk	week(s)
lb	pound(s)		wt	weight
psi	pounds per square inch		yr	year(s)
rpm	revolutions per minute			

Numbers: All numbers used with abbreviations and fractions or decimals are arabic numerals. Table numbers are roman numerals. Otherwise, numbers below ten are to be written out. Numerals should be used for a series (e.g., "0.5, 1, 5, 10, and 20 days"), for pH values, and for temperatures. When a sentence begins with a number, write it out.

Symbols: Special symbols (e.g., Greek letters) must be identified in the margin, e.g.

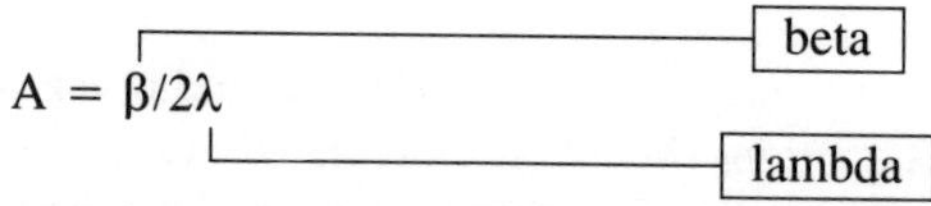

Percent should be % in text, figures, and tables.

Style and format: The following examples illustrate the style and format to be followed (except for abandonment of periods with abbreviation):

Sklarew DS, Girvin DC (1986) Attenuation of polychlorinated biphenyls in soils. Reviews Environ Contam Toxicol 98:1–41.

Yang RHS (1986) The toxicology of methyl ethyl ketone. Residue Reviews 97:19–35.

7. Proofreading scheme

The senior author must return the Master set of page proof to the Editor within one week of receipt. Author corrections should be clearly indicated on proof with ink, and in conformity with the standard "Proofreader's Marks" accompanying each set of proofs. In correcting proof, new or changed words or phrases should be carefully and legibly handprinted (not handwritten) in the margins.

8. Offprints

Senior authors receive 30 complimentary offprints of a published article. Additional offprints may be ordered from the publisher at the time the principal author receives the proof. Order forms for additional offprints will be sent to the senior author along with the page proofs.

9. Page charges

There are no page charges, regardless of length of manuscript. However, the cost of alteration (other than corrections of typesetting errors) attributable to authors' changes in the page proof, in excess of 10% of the original composition cost, will be charged to the authors.

If there are further questions, see any volume of *Reviews of Environmental Contamination and Toxicology* (formerly *Residue Reviews*) or telephone the Editor (see first page for telephone numbers). Volume 98 is especially helpful for style and format.